Value Addition and Processing of Milk and Milk Products

NIPA® GENX ELECTRONIC RESOURCES & SOLUTIONS P. LTD.
New Delhi-110 034

Value Addition and Processing of Milk and Milk Products

Samit Dutta
Principal & Dean
FPT & BE, AAU and Core Co-PI, NAHEP - CAAST
Anand Agricultural University
Anand, Gujarat

Ashish K. Makwana
Associate Professor & Head
Department of Dairy Business Management
SMC College of Dairy Science
Kamdhenu University and Co-PI, NAHEP- CAAST
Anand Agricultural University
Anand, Gujarat

Mahendra D. Gurjar
Assistant Professor, SMC College of Dairy Science
Kamdhenu University and Co-PI, NAHEP- CAAST
Anand Agricultural University
Anand, Gujarat

Vishita Khanna
Assistant Professor
LABMI, AAU and Co-PI, NAHEP- CAAST
Anand Agricultural University, Anand, Gujarat

R.S. Pundir
Principal & Dean
IABMI, AAU and Principal Investigator, NAHEP - CAAST
Anand Agricultural University, Anand, Gujarat

NIPA® GENX ELECTRONIC RESOURCES & SOLUTIONS P. LTD.
New Delhi-110 034

NIPA® GENX ELECTRONIC RESOURCES & SOLUTIONS P. LTD.

101,103, Vikas Surya Plaza, CU Block
L.S.C.Market, Pitam Pura, New Delhi-110 034
Ph : +91 11 4386 0225, 9717133558, 9540816132
E-mail: newindiapublishingagency@gmail.com
Website: www.niparesources.com

Print ISBN: 978-93-58877-39-7

ebook ISBN: 978-93-58872-48-4

Composed and Designed by NIPA®.

ANAND AGRICULTURAL UNIVERSITY
ANAND-388 110, GUJARAT
Tel.: (O) +91-2692-261273, Fax : (O) +91-2692-261520
Email : vc@aau.in

Dr. K.B. Kathiria
Vice Chancellor

Message

Agriculture is backbone of the Indian economy as, inter alia, it provides livelihood to nearly half of the country's workforce. As a result of technological advancement, enabling policy environment and hard work of farming community the country has not only become self sufficient in production of most of the crops and allied activities but also has made significant strides in the export of some Agricultural commodities. There is no denying the fact that focus in agriculture, since the time of green revolution, has been centric to production technologies.

Marketing, the key problem of farmers has not received adequate attention. Agriculture market intelligence an integral part of marketing becomes important and crucial when more than 80 per cent of the farmers in the country, who are small and marginal and constitute core production system, hardly understand the market dynamics. In view of this, there is dire need of rigorous research on various aspects of agricultural marketing with focused attention on market intelligence accompanied with capacity building of PG students, faculties, policy makers and farmers. This may go a long way in improving the existing level of higher education and eventually helping farmers to increase their income.

The Indian Council of Agricultural Research, New Delhi, with the active support of the World Bank has formulated a National Agricultural Higher Education Project (NAHEP), for improving the level of agricultural education and research in the country. In view of this, Center for Agricultural Market Intelligence under Center for Advanced Science and Technology (CAAST) component of NAHEP has been awarded to Anand Agricultural University through a competitive selection process.

As a part of its basic objective the center for agriculture market intelligence at AAU is gainfully engaged in, inter alia, conducting research in the broad area of agriculture marketing and market intelligence and organizing capacity building programmes, online and offline, involving experts of national and international repute for various stakeholders including PG students, faculties and farmers.

I am happy to know that the center has been rated top performer in the category of second phase by the World Bank and ICAR in its mid-term review meeting. I am sure the center will not only achieve its objectives under NAHEP-CAAST but also scale new heights in the time to come.

I congratulate the whole NAHEP CAAST team consisting of faculties, research staff, in general, and members of the research team, in particular of 10 commodity reports covering Wheat, Maize, Groundnut, Cumin, Potato, Poultry, Fisheries, Dairy and Food Processing. I also hope that these commodity reports will be useful for stakeholders, researchers, policy makers and all those interested in the broad area of agriculture market intelligence including allied fields.

Date : 26/04/2021

(K.B. Kathiria)

Acknowledgement

The 'Centre for Agricultural Market Intelligence' under NAHEP-CAAST at AAU, Anand is comprehensive and encompasses all crucial issues, including demand-supply predictions, price forecasting, and capacity building of various stakeholders with a focus on PG students and faculties. This project is a collaborative endeavor from ICAR-World Bank, and we acknowledge the technical and financial support received from ICARNAHEP and World Bank.

This book is prepared by research teams mainly using time series data apart from discussions and valuable suggestions from the experts in the broad area of Agriculture and Agricultural Marketing. The book contains comprehensive observations and findings emanated from the analysis using the best available data and appropriate analytical tools. We expect that this publication will be helpful to all the concerned stakeholders, including P.G. students, researchers, and policymakers.

We are immensely indebted to Dr. R. C. Agrawal, DDG (Edu.) & National Director NAHEP, ICAR, New Delhi, for continuously enlightening us with his valuable guidance, direction, stewardship, and excellent support in the overall implementation of the project, as also bringing out this publication.

We are highly thankful to our Hon. Vice-Chancellor Dr. K. B. Kathiria and former Hon. Vice-Chancellor Dr. R. V. Vyas for their guidance and a keen interest in the timely progress of all project activities, in line with their expectations. Moreover, our Hon. Vice-Chancellor Dr. K. B. Kathiria has played a vital role in this prestigious project as the project was secured by the university when he was Director of Research & Dean P.G. studies. He is primarily responsible for all research projects, including national and international.

We are also thankful to Dr. (Ms) Anuradha Agrawal, National Coordinator (CAAST & Component 2) and Dr. Prabhat Kumar, Former National Coordinator, NAHEP-CAAST, ICAR, New Delhi, for their special interest in the implementation and growth of the project through their significant suggestions and the constant guidance in general and preparing this report in particular. We have also greatly benefitted from their office colleagues and staff, which is highly appreciable.

Helps rendered by the Director of Research & Dean P.G. Studies, Dr. M.K. Jhala, who is inter alia, responsible, in more ways than one, are thankfully acknowledged. His wholehearted support keen interest and professional way of monitoring of all activities pertaining to the project are highly encouraging and inspirational for all of us.

Dr. Y.C. Zala, Former Principal & Dean, International Agribusiness Management Institute and Officer In-charge has been deeply involved in all the activities since the inception of the project. The team thankfully acknowledges the assistance provided by Dr. Y.C. Zala.

We would like to acknowledge untiring efforts, leadership, and continuous guidance provided by Principal Investigator, NAHEP-CAAST, AAU which led to the completion and betterment of this book. We are also thankful for the collective efforts of whole team of NAHEP CAAST.

Place : NAHEP-CAAST, AAU, Anand **Research Team**

Preface

The agricultural market intelligence project under National Agricultural Higher Education Project of the Indian Council of Agriculture Research is a unique effort to encompass the relationship between milk producers, milk supply chain, organized and unorganized dairy processors, and consumers. It is one of its kind of research where a team of self-motivated researchers has come up together to resolve some key issues related to the agriculture sector under the guidance of the agricultural university.

The main aim of this market intelligence project is to increase farmer's income by giving them insight into the market and improve the decision-making ability. Along with this, the project also aims to providing farmers with the potential to obtain quantifiable information to continuously measure and monitor farm operations and react accordingly. Farmers who can complement these technologies with suitable operational and commercial practices could see profitability increase significantly.

Global milk production is around 852 million tons and the demand for dairy is largely driven by increasing urbanisation and rising incomes in emerging markets. The global dairy sector is currently going through a period of turbulence. Slowing demand from China, Russia's trade embargo and the removal of EU milk quotas, resulted in a period of excess supply and low prices. Despite this, the long term outlook for the sector remains positive. Rising populations and changing diets are increasing demand for dairy. as opposed to basic carbohydrates (mostly grains).

Dairy is convenient and relatively cheap source of protein. India is the largest producer and consumer of milk and milk products, with a production of 187 million tons milk in year 2019. In India, about 46 per cent of the milk produced is either consumed at the producer level or sold to non-producers in the rural area, the balance 54 per cent of the milk is available for sale to organized and unorganized players. The organized sector comprises of Government, Producers' Owned Institutions, and Private players. The unorganized/ informal sector involves local milkmen, dudhias, contractors, etc. The Indian consumption pattern of dairy product is basically inclined towards consumption of traditional products although, with the increase in globalisation the trend of western products is also gaining momentum in the urban. The milk production in Gujarat has reached a volume of 14.96 million tons in 2019. Gujarat ranks fifth in dairy markets of India and the major milk contribution comes from cow and buffalo milk.

The 'Traditional' value drivers of price, taste and convenience have been supplemented by 'Evolving' drivers of health and wellness, safety, social impact and experience. Overarching these drivers is a desire for transparency from food companies. Evolving consumer trends such as the demand for healthy and 'clean label' products, increased demand for 'functional foods' and the growth in protein consumption will drive innovation in the sector. While having strong technological capabilities is key to taking advantage of these trends, it is not sufficient.

Though the dairy sector is currently dealing with a number of challenges, favourable long term consumer trends and developing technologies provide a number of opportunities. This report looks at these challenges and opportunities at the producer, processor and retailer levels. It then outlines some of the overarching themes presented by these challenges and opportunities, and looks at how the dairy industry as a whole needs to adapt.

As of date, the Agricultural Market Intelligence Center, Anand Agricultural University, Anand has carried out three lectures under the theme of food processing & value-addition, three webinars on Impact of COVID – 19, Trade in Fruit & Vegetables Products and Dairy Commodities, and Value-addition, along with one workshop on 'Methodology for price forecasting, market competitiveness, and export opportunities assessment for dairy & food products' under the capacity building component of the project with huge participation from post-graduate students, research scholars, faculty members and other stakeholders across the country. In general, about 1500 registrations were recorded in the organized lecture series while around 2500 registrations were recorded in the webinars.

The Food & Dairy group of this NAHEP projet is currently working on two research projects viz., 'Growth and Export opportunities of major value-added functional food, health food, and traditional Indian dairy products', and 'A study on Market Competitiveness of Dairy Commodities'. As of date, the Secondary data analysis for each selected product for the aforesaid research has been concluded to examine the growth, trade direction, instability, comparative advantage, and nominal protection coefficient. To examine the challenges and opportunities in the export of these products, a Survey of the concerned stakeholders will be carried out for primary data collection.

This informative report attempts to present a comprehensive scenario of value-added milk products and dairy commodities at the global, national, and state (Gujarat) level. From 55.5 million tons milk Production in 1991-92 to 187.7 million tons milk production in 2018-19 India has witnessed compound growth of 4.44 per cent. An Attempt has been made to give a detailed insight on organized and unorganized dairy sector, marketing and distribution of milk and milk products, trade performance of Indian dairy sector, various value-addition concepts for milk products, Impact of COVID pandemic on Indian dairy sector, government schemes and policies for dairy entrepreneurship, and challenges that lie within the dairy industry as well as awaiting opportunities of the sector to engage and encourage the youth to participate in strengthening the sector.

We sincerely wish that this commodity report would prove beneficial to different stakeholders including students, research scholars, farmers, faculties, and Industries.

Authors
NAHEP-CAAST, AAU, Anand

Contents

Abbreviations

AMCU	: Automatic Milk Collection Unit
BCD	: Basic Customs Duty
BIS	: Bureau of Indian Standards
BMC	: Bulk Milk Cooler
CAGR	: Compound Annual Growth Rate
CAPEX	: Capital Expenditure
CEPA	: Comprehensive Economic Partnership Agreement
CIF	: Cost Insurance and Freight
DAHD	: Department of Animal Husbandry & Dairying
DCS	: Dairy Co-operative Society
DHANI	: Dairy Health and Nutrition Initiative
DMS	: Delhi Milk Scheme
DOAH	: Directorate of Animal Husbandry
DPMCU	: Data Processor Milk Collection Unit
EIA	: End Implementing Agencies
FAO	: Food & Agriculture Organization
FICCI	: Federation of Indian Chambers of Commerce and Industry
FOB	: Free on Board
FPO	: Farmer Producer Organization
FSSAI	: Food Safety and Standards Authority of India
FTA	: Free Trade Agreement
GCMMF	: Gujarat Co-operative Milk Marketing Federation
GDP	: Gross Domestic Product
GVA	: Gross Value Added
HoReCa	: Hotel Restaurant and Catering
KCMMF	: Kerala Co-operative Milk Marketing Federation
LLPD	: Lakh Liters Per Day
MEIS	: Merchandise Export from India Scheme
MMT	: Million Metric Tones
MNC	: Multinational Company
MPC	: Milk Producer Company
MT	: Million Tones
NABARD	: National Bank for Agriculture & Rural Development
NCDC	: National Centre for Disease Control
NCDEX	: National Commodities & Derivatives Exchange
NCDFI	: National Co-operative Dairy Federation of India Ltd
NDP	: National Dairy Plan
NDS	: NDDB Dairy Services
NeML	: NCDEX E-Market Limited
NFDM	: Non-Fat Dried Milk
NRLM	: National/State Rural Livelihood Mission
QSR	: Quick Service Restaurant
RCEP	: Regional Comprehensive Economic Partnership

RKVY : Rashtriya Krishi Vikas Yojana
RoSCTL : Rebate of State and Central Levies and Taxes
RUC : Ready to Use Culture
SCF : Super Critical Fluid
SMP : Skimmed Milk Powder
SNF : Solid Not Fat
UHT : Ultra-High Temperature
USDA : United States Department of Agriculture
USP : Unique Selling Proposition
WMP : Whole Milk Powder
WPI : Wholesale Price Index
WSP : Warehouse Service Provider
WTO : World Trade Organization

1

Dairy Sector – Overview

1.1. Global Dairy Scenario

Global milk production is forecast at 852 million tonnes in 2019, an increase of 1.4 per cent from 2018 (DAHD, 2019) – a smaller rate of growth, reflecting downward revisions made for India and the European Union. Much of the anticipated output expansion will originate in India, Pakistan, China, the European Union, and Brazil, partially offset by declines in some countries including Australia, Colombia, and Argentina. In India and Pakistan, herd expansions drive output growth, while in China, farm efficiency improvements underpin higher growth. The output is increasing in the European Union, although slowly as dry weather limited the supply of milk during the summer, whilst rising dairy herds and stable milk prices are supporting higher production in Brazil. By contrast, the output may decline in Australia and Colombia due to dry weather, whereas in Argentina, rising feed costs and restrained consumer demand may dampen production. Elsewhere, in the United States, higher milk yields hold up the growth momentum, whereas, in New Zealand, favourable weather corroborates a positive production outlook (DAHD,19).

The dairy industry has an important role to play in the global food system, and the change in consumer lifestyle has increased demand for dairy products across the globe. This change in the consumption pattern has resulted in boosting the scope and intensity of the global dairy trade. Industries across the globe face strategic challenges because of globalization and so does the dairy industry. Although the countries in South Asia largely rely on dairy farming for their livelihood, and India is the world leader in milk production, the industry is highly unorganized and primarily managed as family businesses. Being mostly family managed, the Indian farmers ensure low cost in production, making their products competitive in the global market. Barring a few national players, such as Amul and Mother Dairy, the industry is highly informal and there is the scope for multi-fold improvement.

As shown in the Figure 1.1, world milk production has reached 843 million tonnes in 2018 which was 530 million tonnes in 1988. India contributes 22 per cent in the global production followed by USA, China, Pakistan, and Brazil. The countries with highest milk surplus are New Zealand, USA, Germany, France, and Australia. On the other hand, there are countries like China, Italy, Russian Federation, Mexico, and Indonesia that have highest milk deficit (FAO, 2018).

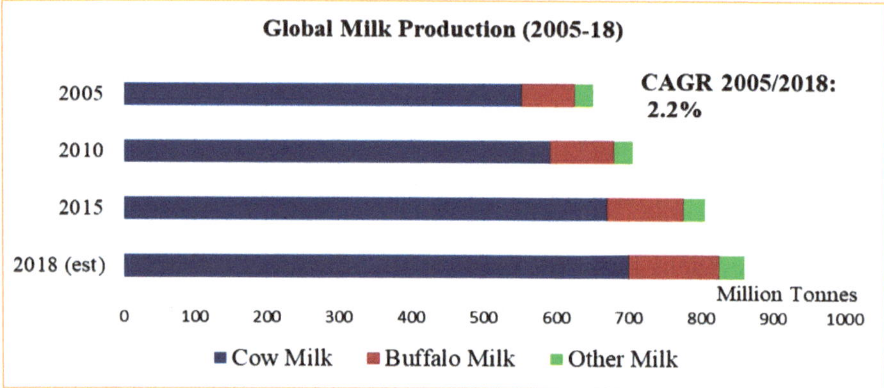

Fig. 1.1: Global Milk Production
Source: FAO, 2018

1.2. Indian Dairy Scenario

1.2.1. Population & Composition of Milch Animals

Between the two Indian Livestock Censuses (2012 and 2019), the relative share of cow's milk versus buffalo milk grew three per cent to 48 per cent due to an increase in milk contribution from exotic and crossbred cows. A predicated 49 per cent of India's milk production originates from water buffalo (35 %indigenous buffalos and 14 % non-descript buffalos (ones which are not selected or bred for productivity). Uttar Pradesh has the heaviest massing of buffalo population at 33 per cent, followed by Rajasthan (12.4%), and Madhya Pradesh (9.5%) (DAHD, 2019).

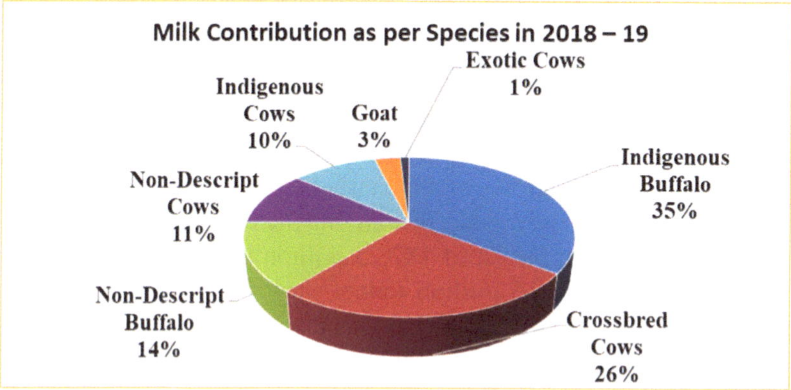

Fig. 1.2: Milk Contribution as per Species in 2018 – 19
Source: DAHD, 2019

Figure 1.2 shows the contribution of milk production by Cow, Buffalo, and Goat. Further, Contribution from Cow is divided into four major categories namely, Crossbred, Non-Descript, Indigenous and Exotic and their share is 26, 11, 10 and 1 per cent respectively. Similar to cows, water buffaloes are also divided into two major categories that are Indigenous Buffalo that contribute the highest i.e., 35 per cent and Non- Descript Buffalo (i.e., animals not selected or bred for productivity) contributing 14 per cent to the total milk production. The remaining 3 per cent is contributed by Goat milk (DAHD, 2019).

As far as livestock population is considered, Uttar Pradesh has the heaviest concentration of buffalo population (33 %) followed by Rajasthan (12.4 %) and Madhya Pradesh (9.5 %). Although the share of goat milk production is only three per cent but is reportedly growing in states like Rajasthan (13.8 %) and West Bengal (10.9 %) as the unproductive ones can be sold to slaughterhouse which makes it commercially viable. Farmers prefer water buffaloes because their milk holds a higher fat content (7-8 %), which often receive higher market prices as milk prices are set on volume, fat, and solidsnot-fat (SNF) content.

The livestock sub-sector plays a vital role in the Indian economy and also in the socio-economic development of millions of rural households. In rural areas, livestock is a major source of animal grazing and offers milk, meat, eggs, wool, hides & skins, manure, and fuel. It accounted for about 4.9 per cent of the total GDP/GVA of India (at current prices) and 28.4 per cent of the GDP/GVA of the Agriculture & allied sector. Although the contribution of agriculture & allied sector to the national GDP has declined during the past few decades, the contribution of the livestock sub-sector has remained stable. In real terms, the contribution of agriculture & allied sector to the national GDP has been falling from a level of 23 per cent in 1999-20 to 17 per cent in

2017-18. However, the contribution of livestock to agriculture & allied sector gross domestic product (GDP) has increased from 23.12 per cent to 28.4 per cent during the same period (DAHD, 2019).

1.2.2. Milk Production

The Dairy sector in India has grown significantly over the years. As a result of prudent policy intervention, India ranks first among the world's milk producing nations, achieving an annual output of 187.75 million tonnes during the year 2018-19 as compared to 176.35 million tonnes during 2017-18 recording a growth rate of 5.33 per cent (DAHD, 2019). This represents a sustained growth within the supply of milk and milk products for the growing population. India is the world's largest producer and consumer of milk accounts for 22 per cent of the world's total milk production (DAHD, 2019). Dairy manufacturing is one of India's leading industries, the second most vital segment of the food-processing sector.

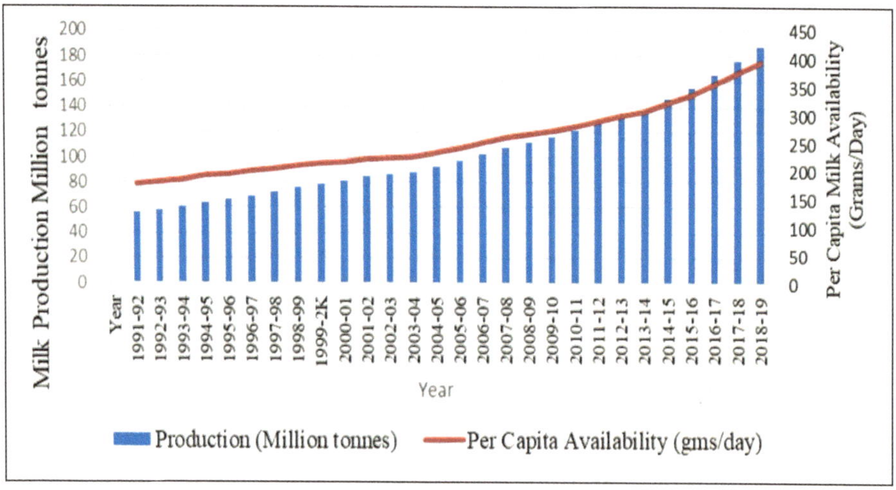

Fig. 1.3: Milk Production & Per capita Milk Availability in India
Source: Department of Animal Husbandry and Dairying. (2020).

Dairying has become an important secondary source of income for millions of rural families and has presumed the most important role in providing employment and income generating scope particularly for women and marginal farmers.

Milk & milk products constitute a major share of the value of output from the livestock subsector; their share increased from less than 50 per cent in 1950-51 to 67 per cent in 2017-18 (at current prices) (DAHD, 2019). Milk has become the single largest agricultural commodity in terms of the value of the commodity produced. The value of milk output in 2017-18 was Rs. 7,01,530 Crore – surpassing the total value of output from food grains (DAHD, 2019).

Table 1.1: Major Indian Milk Producing States & Cattle Yield

Sr. No	States	Total Milk Production MMT	Yield (Nondescript/ indigenous Cow) Kg/Day	Yield (Exotic/ crossbreed) Kg/Day	Yield Buffalo Kg/Day
1	Uttar Pradesh	30.519	2.99	7.15	4.44
2	Rajasthan	23.668	4.32	7.74	6.16
3	Madhya Pradesh	15.911	2.71	8.35	4.31
4	Andhra Pradesh	15.044	2.67	7.78	5.65
5	Gujarat	14.493	4.22	8.96	4.95

Source: APEDA,2020 & DAHD, 2017, (Cattle and Buffalo Milk Yield Data Shown here is for the Year 2016-17)

The increasing importance of dairy to the Indian economy means the international competitiveness of the industry is on the rise. More importantly, globalization and trade liberalization coupled with the easy flow of information and advancement in transportation and communication technology have resulted in an unprecedented intensification of market competition worldwide. The analysis of worldwide scenarios of the dairy industry may help us to understand the position of the Indian dairy sector concerning its global competitors. If the degree of worldwide competition increases, then it's going to become a threat to an uncompetitive economy. On the opposite hand, if it decreases and even though the market share of the country under consideration is static, it may be an opportunity.

India continues to be the enormous producer of milk in the world. Several initiatives have been introduced by the government to improve livestock productivity, this resulted in significant growth in milk production from 102.6 million tonnes at the end of the Tenth Plan (2006-07) to 191 million tonnes in 2019. During 2017-18 and 2018-19, milk production was 176.3 million tonnes and 187.7 million tonnes, respectively showing a 6.47 per cent annual growth. (DAHD,2019) The production of milk and corresponding growth rate (%) per year from 2009-10 to 2018-19 is shown in the Figure below.

Low Cost of Milk Production

Milk production in India is not competitive to production of food-grain as is in the developed countries. The land for fodder production has remained stable between 4 to 5 per cent of arable land. It is not likely to increase in the immediate future. Milk production in Punjab and Haryana is likely to depend on cultivable fodder. In rest of India, milk is largely produced from Agro-by-products. The cattle and buffalo population has been growing at a much smaller rate, even though there is need to attain zero growth. Most of the milk in India is produced by otherwise unproductive labour. Therefore, cost of milk production in India is cheapest in the world-perhaps next to New Zealand (Mondal, 2012).

Table 1.2.: Top 20 Countries with Highest Milk Yield per Cow

World Rank	Country	Average liters per cow
1	Saudi Arabia	10,133
2	Israel	10,035
3	Republic of Korea	9,816
4	United States of America	9,314
5	Denmark	8,389
6	Sweden	8,144
7	Canada	7,963
8	Finland	7,873
9	Japan	7,284
10	Spain	7,278
11	Netherlands	7,277
12	United Kingdom	7,271
13	Luxembourg	7,002
14	Czech Republic	6,884
15	Germany	6,877
16	Estonia	6,780
17	Switzerland	6,651
18	Hungary	6,596
19	Jordan	6,521
20	Kuwait	6,448

Source: CIWF, 2012

Although the production cost of milk in India is very less but average milk yield of Indian cattle is much less i.e., 987 kg /year compare to 6273 kg/ year in Denmark, 5289 kg/ year in France, 5938 kg/year in Canada, 5462 kg/ year in United Kingdom, 7038 kg/year in USA and 11000 kg/year in Israel (Bhagyalaxmi,2020). Which indeed is one of the drawbacks for Indian dairy sector, which ranks first in the production of milk but yet not in the top 15 countries list in terms of milk productivity. This high yield could be achieved through proper feed, water management and housing. Animal productivity is highly dependent on the availability of green feed and fodder, which is a major challenge these days in India. In dry fodder, green fodder, and concentrated goods, India can face substantial deficits. Just four per cent of the crop area is actually allocated to the production of fodder. Owing to competing uses of agricultural land for food and other cash crops, the number might not increase in the short term.

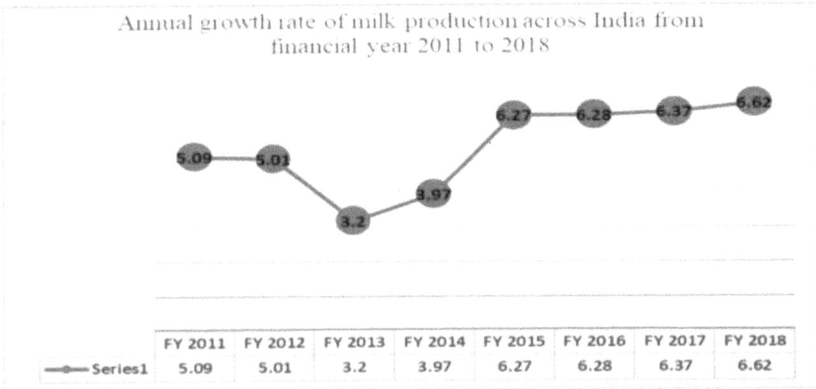

Fig. 1.4: Annual growth rate of milk production across India from financial year 2011 to 2018
Source: Department of Animal Husbandry & Dairying, 2019

Some 80 million rural households, the majority of whom are small marginal farmers and landless, are engaged in milk production as a source of their livelihoods. In dairy farming, start-up costs are lower for farmers and cash flow is steadier than with seasonal planted crops. However, unlike the larger herd sizes of leading milk-producing countries in the world, some 95 per cent of milk producers in India hold just 1 to 5 milch animals per household, which makes this little more than a subsistence-level farming system Dairy farms sized from 50 to 200 cattle are increasing in some of the major dairy states such as Punjab, Gujarat, Maharashtra, and Telangana/Andhra Pradesh, although they are still few.

According to data available from DAHD, GOI, Uttar Pradesh (16.3 %), Rajasthan (12.6 %), Madhya Pradesh (8.5 %), Andhra Pradesh (8.5 %), Gujarat (7.7 %) and Punjab are the top six milk-producing states (6.7 %). In the Indian Fiscal Year 2018/2019, they together accounted for 60 per cent of total milk production (DAHD, 19).

1.2.3. Milk Productivity

Productivity of cows vary largely from country to country as it is dependent on breeding practices, feeding and production systems. The global average productivity of dairy cows in 2018 was around 2,500 litres/cow/year as compared to Indian Average which is 1460 litres/cow/year (FAO, 2018). Key factors responsible for low productivity in India include the low genetic potential of Indian bovines, a lack of nutritious and balanced feed rations (especially green fodder availability), and inadequate veterinary services. Per the Department of Animal Husbandry and Dairying's (DAHD) latest annual report (2019-20), the average milk yield by volume per day in India includes indigenous cattle at 3.85 kilograms (kg) per day, crossbreeds at 7.85 kg, exotic cattle at 11.67 kg, indigenous water buffalo at 6.34 kg, non-descript buffalo at 4.35 kg and goat at 0.45 kg.

As mentioned earlier, animal productivity is vastly dependent on green feed and fodder availability which is a major challenge in India these days. According to the Government of India (GOI), by 2020, India may face significant deficits in dry fodder, green fodder, and concentrates to the extent of 11, 35, and 45 per cent against an estimated demand of 468 MMT, 213MMT, and 81 MMT, respectively. Currently, only four per cent of the cropping area is dedicated to fodder cultivation. That number may not increase in the short term due to competing uses of agricultural land for food and other cash crops. As such, compound cattle feed use is just 8-10 MMT against India's total dairy feed requirement of around 80 MMT (DAHD, 2019).

Average Yield Rate for Milk: The average yield of milk per day per animal in milk at the National level from different species during 2018-19 is given below:

Table 1.3: Average Yield Rate of Milk

Exotic Cows (kg/day)	Crossbreed Cows (kg/day)	Indigenous Cows (kg/day)	Non-Descript Cows (kg/day)	Indigenous Buffalo (kg/day)	Non-Descript Buffalo (kg/day)
11.67	7.85	3.85	2.50	6.34	4.35

Source: DAHD, 2019

1.2.4. Milk Consumption

Milk and Milk products have a share of 8 per cent of consumer expenditure in rural India and 7 per cent in urban India in 2011-12 (MOSPI,2012).

Table 1.4: Consumption of liquid milk :2004-05 to 2011-12

Year	Per capita consumption (Liters)		Per cent of consuming households (in 30 days)	
	R	U	R	U
2004-05	3.866	5.107	71.3	85
2009-10	4.117	5.358	76.4	84.9
2011-12	4.333	5.422	78	84.9

Source: MOSPI, 2012

Between 2004-05 and 2011-12, estimated per capita consumption of liquid milk saw a rise of 470 ml per month in rural India and 315 ml per month in urban India.

Based on the milk and dairy product consumption patterns across the 29 states and seven union territories, the top five states that consumed 50 per cent of dairy products in 2019 were Uttar Pradesh (19 %, highest share), Rajasthan (9 %), Gujarat (8 %), Maharashtra (7 %) and Bihar (7 %). On the supply side, the top five milk-producing states were Uttar Pradesh (16.3 %), Rajasthan (12.6 %), Madhya Pradesh (8.5 %), Andhra Pradesh (8.5 %), and Gujarat (7.7 %). Together, they accounted for 53.6 per cent of total milk production in IFY 2018/2019 (USDA GAIN, 2020).

1.2.5. Milk and milk products

Over the past decade, milk processing capacity has increased due to rise in the demand of good quality milk and milk products that are made with proper sanitation and packaged to make the product aesthetic. Out of the total milk produced in India, 65 per cent to 70 per cent is sold in raw form. The rest gets converted into products like ghee, cheese, butter, yogurt, milk, dry milk powder, casein, and dessert, using processes like chilling, pasteurization, and homogenization. Distinctive by-products include the following: buttermilk, whey, and their derivatives.

Table 1.5: Utilization of milk in different dairy products

Items	**Percentage in relation to**	
	Total Milk Production	**Total Quantity converted into milk products**
Fluid milk	44.5	-
Manufactured milk	55.5	(100.0)
Ghee	32.7	58.9
Dahi	7.8	14.0
Butter	6.3	11.4
Khoa	4.9	8.8
Ice-cream	0.7	1.3
Cream	1.9	3.4
Other Products	1.2	2.2

Source: Indiaagronet

In India, both the cooperative and private sectors have played a notable role in milk processing. As shown in the below figure 1.5.

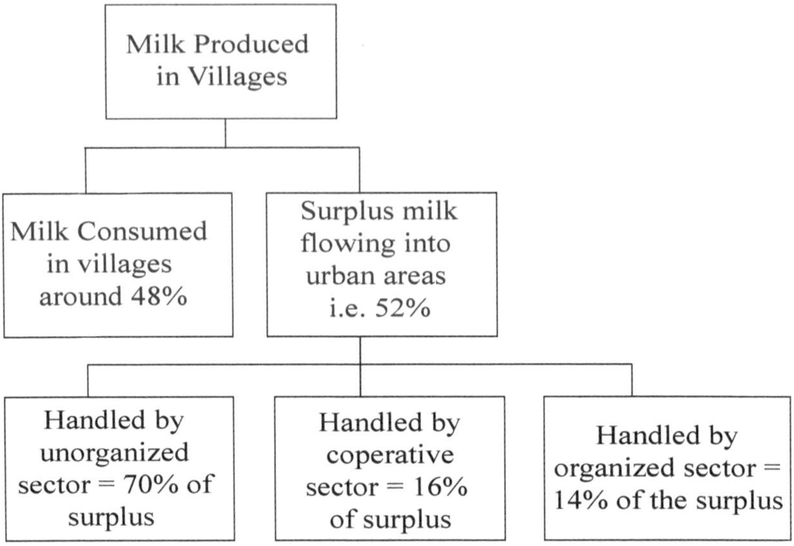

Fig. 1.5: Distribution and Consumption pattern of milk produced in villages
Source: National Dairy Plan 2007-08 to 2011-12

In India, there are various brands dealing in Dairy products like Mother Dairy, Aarey and DMS. Aarey caters to Mumbai and DMS caters to Delhi. Private sector brands are Nestle, Britannia Milkman, Arokya, Metro Dairy and Paras. In addition to the prominent brands, there are a large number of small regional brands operating in different cities. By one estimate, the small brands have about 40 per cent share of the market as mentioned in the below Table 1.6.

Table 1.6: Processing Capacities of Major Dairy Processors in India

Companies	Products	Processing Capacity per day LLPD
Amul	Packaged Milk, Curd, Buttermilk, traditional sweets, Ghee, Cottage cheese, Ice- cream	230
Aavin	Curd, Lassi, Paneer, Yoghurt, Traditional sweets, Probiotic and Fermented dairy products	40
Sterling Agro	Butter, Butter oil, Ghee, Packaged milk, SMP, WMP, Dairy Whiteners, Flavoured milk, Cottage Cheese, Curd and Buttermilk	26
Verka	Packaged milk, Yoghurt, Ghee, Butter, Ice cream, Cheese and Traditional Sweets	20
Creamline Dairy Products (Jersey)	Packaged Milk, Curd, Buttermilk, traditional sweets, Ghee, Cottage cheese, Ice- cream	10
Vijaya	Packaged Milk, Ghee, Sweets, Curd	8

Companies	Products	Processing Capacity per day LLPD
Britannia	Cheese, Dahi, Milk Based Beverages, Dairy Whitener, Packaged milk	4
Kwality	Milk, Ghee, Butter, Milk powder, Curd, Yogurt, Cheese, Ice cream	3
Vadilal	Ice cream, Flavored milk, Paneer, Ghee	2
Nestle	Packaged milk, Curd, Yogurt, milkmaid, Ghee, Chocolates	1.5
Mother Dairy	Packaged Milk, Dahi, Lassi, Misti Doi, Flavoured milk liquid and powder, cheese, butter, paneer, yoghurt, traditional sweets, Ice cream	1
Dynamix Dairy	Butter, Ghee, SMP, WMP, Dairy Whiteners, Flavoured milk, Cottage Cheese, Curd	1
Kraft	Cheese, Butter, Ghee	1
Haryana milk foods	Ghee and milk powder	0.25

Source: Company Websites

The Indian consumption pattern of dairy product is basically inclined towards consumption of traditional products although, with the increase in globalisation the trend of western products is also gaining momentum in the urban. Indian products also have a very high demand in the export market such as casein for food processing or pharmaceuticals. The major export destinations for the Indian dairy products are Bangladesh, Middle East, US and Egypt. Indian Dairy Market by Commodities In terms of products, skim milk powder, Casein and ghee are the most important product exported from India followed by butter and whole milk powder.

1.3 Milk production in Gujarat State

The geographic area of Gujarat is 196244 Sq. Km, which is 5.97 per cent of the country's geographic area. The state of Gujarat consists of 33 districts, 250 Talukas, 18225 villages (including uninhabited villages). The milk production in Gujarat has reached a volume of 15.4 Billion Litres in 2019. Gujarat ranks fifth in dairy markets of India and the major milk contribution comes from cow and buffalo milk with a growth rate of 13 per cent (CAGR) during 2014-19 (NDDB,2019)

1.3.1 Livestock Census of Gujarat State

According to the Livestock Census- 2019, the total livestock population in the State was 268.71 lakh in 2019, registering a decrease of 0.95 per cent over the year 2012. Out of the total livestock reported in Census- 2019, cattle, buffalo, sheep goats and others were 96.11 lakhs (35.77 %), 105.43 lakhs (39.24 %), 17.87 lakhs (6.65 %), 48.68 lakhs (18.12 %) and 0.61 lakhs (0.23 %) respectively. The percentage increase in the total population of buffalo and sheep was 1.52 per cent, 4.66 per cent respectively over the previous Census-2012. The percentage decrease in the total population of cow and goat was 3.74 per cent and 1.84 per cent respectively in comparison to the previous Census-2012.

Gujarat is fortunate that it has healthy and high-yielding cattle and buffalo breeds. Mahesani, Jafarabadi, Banni, and Surti are well known for their high capacity to yield milk (DAHD, 2019).

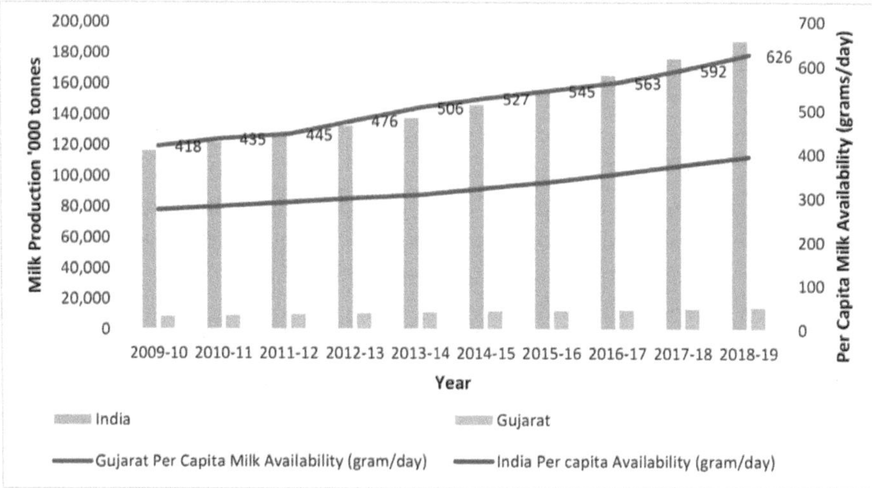

Fig. 1.6: Growth in Milk Production & Per Capita Milk Availability
Source: DOAH, 2019, GoG

India is the largest milk-producing country in the world, which is a prestigious matter for the country. Among the estimates of total milk production of the country, Gujarat state gives about 7.72 per cent contribution and also enjoys the fifth rank among the all state and union territory of the country in the year 2018-19. Gujarat state had contributed 1.75 per cent milk towards the whole world's milk production in the year 2017. This for the state is a special milestone.

In the year 2018-19 estimated milk yield per day per in milk animal of various category viz. crossbred cow, indigenous cow, Non-Descript Cow, Indigenous Buffalo, Non-Descript Buffalo and goat of the Gujarat state is 9.32 kgs,4.65 kgs, 3.84 kgs.,5.39 kgs,4.44 kgs. and

0.48 kgs. respectively. Whereas milk yield per day of India is 7.85 kgs., 3.85 kgs. 2.50 kgs. 6.34 kgs. 4.35 kgs. and 0.45 kgs. respectively, this shows high productivity as compared to the national average (Except indigenous Buffalo).

1.3.2. Composition of Milch Animals in Gujarat

The total estimated milk productio n for the state during 2018-19 works out to be 14492.40 thousand tonnes, which shows an increase of 6.80 per cent over the previous year's estimate of 13569.13 thousand tonnes. Estimates of milk production by species are given in table – 1.7 with percentage variation over the previous year. Percentage contributions of milk production by species to the total milk production for the year 2017-18 and 2018-19 are as under.

Table: 1.7: Percentage contributions of milk production by species to the total milk production

Sr.No	Species	Milk Contribution (%)	
		2017	2018
1.	Cow		
	Indigenous	22.69	22.14
	Crossbreed	23.90	25.57
	Total Cow	46.59	47.72
2.	Buffalo	51.09	50.08
3.	Goat	2.32	2.20
	Total (per cent)	**100.00**	**100.00**

Source: DOAH, 2019, GoG

2

Marketing & Distribution of Milk & Milk Products

In India, about 46 per cent of the milk produced is either consumed at the producer level or sold to non-producers in the rural area, the balance 54 per cent of the milk is available for sale to organized and unorganized players. The organized sector comprises of Government, Producers' Owned Institutions (Milk Cooperatives & Producer Companies), and Private players which provides a fair and transparent system of milk collection round the year at the village level. The unorganized/informal sector involves local milkmen, dudhias, contractors, etc. and they are mostly found to be opportunistic, as there is no uniformity of milk price paid to producers and it varies depending upon the situation. The possibility of adulteration of milk is higher among these unorganized groups. In the areas where competition is high and the presence of the formal sector is strong, the unorganized sector generally gives higher prices and at the same time, the unorganized sector doesn't offer remunerative prices to the producers where the organized sector is not present.

Presently, India's total installed milk processing capacity is 66.3 million liters/day in Indian dairy cooperatives, 73.3 million liters/day in private sector companies, and 2.5 million liters/day at producer companies (USDA GAIN, 2020). Unlike the unorganized milk sector, these companies have wide procurement and distribution networks, which include village-level milk collection and chilling centers. The milk collected at these centers is processed in dairy plants, which involves pasteurization, standardization, branding, packaging, and preparation of certain value-added products. Of the total Indian milk production, an estimated 48 per cent is either consumed at the producer level or sold to small outlets in the rural area; the remaining 52 per cent is processed and made available for sale to consumers in urban areas. (USDA GAIN, 2020).

Among dairy products, the market for packaged milk is one of the faster-growing segments, followed by various value-added dairy products. It is marketed as pasteurized milk, either as a mix of cow and buffalo milk or in pure form, in various percentages of fat content. Concerns about milk safety and quality drive strong demand for packaged, ultra-high temperature (UHT) milk, which is sold in aseptic packaging which gives a longer shelf life. By 2022, from 20-21 per cent to 41 per cent of total milk production, the GOI's National Action Plan for Dairy Development aims to double organized milk production. Milk handling (production) by cooperatives has been targeted to increase from 10 to 20 per cent and the private sector from 10 to 30 per cent. If achieved, it would give dairy farmers greater access to the organized milk-processing sector and ultimately greater incomes.

2.1. Demand

The drivers of demand for milk in India are - population growth, urbanization, and increasing per capita income. The consumption of milk has been rising, commensurate with the increase in purchasing power of people, changing food habits & life-styles, and demographic growth. Milk with its varied benefits is that the only source of animal protein for the largely vegetarian population of the country. Also, this rise is driven by factors such as increased consumer interest in high protein diets and raising awareness and availability of dairy products across channels such as an organized retail chain.

The milk consuming population has been consistently rising in the country both in rural and urban areas. According to the Consumer Expenditure Survey (CES, 2011-12) of NSSO, about 78 per cent and 85 per cent of the rural and urban population respectively reported consumption of milk in the country. The growth in the above factors indicates that the demand for milk & milk products will rise consistently in the future.

The below table shows the data of Country's Average Expenditure on Milk and Milk Products as per the data available from NSSO. The table shows that although availability of milk and milk products is better in Gujarat still the expenditure on monthly basis is less compared to Himachal Pradesh and Rajasthan.

Table 2.1: Category-Wise Average Monthly Expense on Dairy

Average Expenditure (INR/Month) on Milk/ Milk Products			
	Tribal	**SC/OBC**	**General**
Himachal Pradesh	165.68	163.13	234.28
Rajasthan	173.29	204.19	269.43
West Bengal	42.48	47.43	62.32
Jharkhand	60.46	68.73	145.5
Orissa	34	42.96	66.68
Chhattisgarh	41.87	46.93	97.63
Madhya Pradesh	57.03	105.94	165.49

Average Expenditure (INR/Month) on Milk/ Milk Products			
	Tribal	**SC/OBC**	**General**
Gujarat	117.31	157.42	226.2
Maharashtra	63.79	83.68	148.75
Andhra Pradesh	63.86	83.61	131.55
Karnataka	60.71	73.34	106.74
Combined Sample	85.06	105.66	161.32

Source: MOSPI,2017

Organized Dairy Sector

2.2. Co-Operative Sector

At present, 222 dairy cooperative milk unions have covered about 17.01 million farmers under the ambit of 1,94,007 village level dairy corporative societies up to December 2019. During the year 2019-20 up to December 2019, the Cooperative Milk Unions procured an average of 464.34 lakh Kg per day of milk compared to 494.27 kg per day during the same time in the previous year, recording a 6.1 per cent decline. The sale of liquid milk by the Cooperative Dairies has reached 370.04 lakh liters per day during the year 2019-20 up to December 2019 as compared to 353.05 lakh liters per day during last year, an increase of 4.8 per cent.

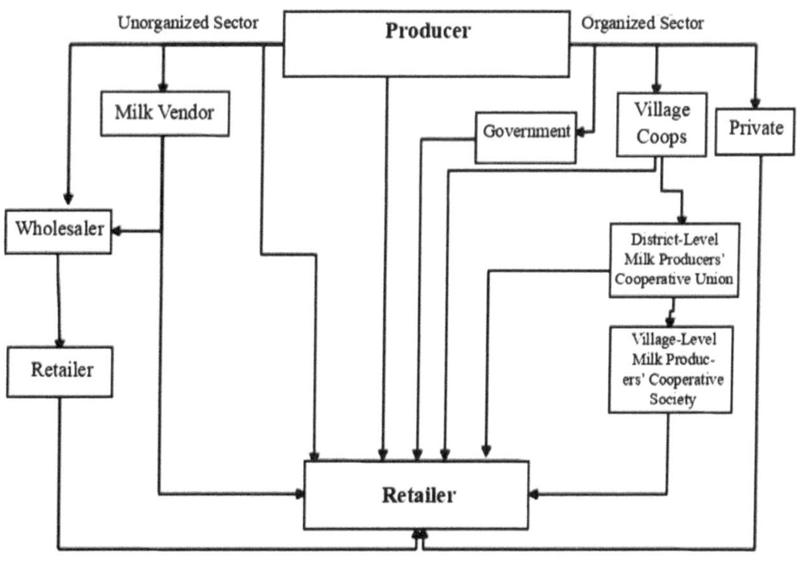

Fig. 2.1: Milk Marketing Channels in India
Source: Rajendran & Mohanty, 2004

Three-tier structure

Village Cooperative Society: The main pattern followed by Village Cooperative Societies in India is an Anand model village dairy cooperative society (DCS) of milk producers. By purchasing a share and committing to selling milk only to society, any producer can become a DCS member. Each DCS has a milk collection centre where milk is taken daily by members. The consistency of milk of each member is checked with payments based on the percentage of fat and Solids-Not Fat (SNF). A part of the DCS income is used at the end of each year to pay each member a patronage incentive dependent on the quantity of milk poured.

The District Union: The Union of District Cooperative Milk Producers is controlled by dairy cooperatives. The Union purchases milk from all societies then processes and markets liquid milk and products. Most Unions also provide DCSs and their members with many inputs and services: feed, veterinary care, artificial insemination to support milk production growth, and the business of the cooperatives. Employees of the Union train and offer advisory services to assist DCS leaders and workers.

The State Federation: The cooperative milk producers' unions in a state form, a State Federation, which is responsible for marketing the fluid milk and products of member unions. Some federations also manufacture feed and support other union activities.

Double Axis Pricing System

In India, in the organized milk market, the double-axis or two-axis milk pricing method is commonly followed, as first recommended by the Royal Commission on Agriculture, formed by the Government of India in 1919 and re-suggested by the Milk Pricing Committee in 1972. The two-axis pricing strategy for setting milk procurement prices based on fat content and SNF has been considered scientifically rational. The minimum SNF content desired for cow milk is fixed at 8.5 per cent and that for buffalo milk is 9 per cent for accepting milk at the collection centers of these dairies and the price is mainly determined by the percentage fat content of the milk. In the organized dairy sector in India, rewards for higher quality milk and penalties for not meeting appropriate minimum standards are prevalent.

Two milk sources are a major issue in the pricing of milk procurement using this system viz. Milk from cows and buffaloes. Because of the fat and SNF content in cow and buffalo milk, this issue has arisen since buffalo milk contains a higher percentage of fat whereas cow milk contains low fat and buffalo milk prices are always higher than cow milk prices. Most dairy plants have some sort of purchasing strategy, which is related to what the plants get from their milk product sales. Varied payment systems and rates have been implemented by various state governments. Some pay less for cow's milk; some pay the same purchasing price for both milk and take a calculated risk that some buffalo milk adulteration will be passed on as cow's milk. Generally, food producers pay solely based on fat.

Status of Dairy Co-operatives in India (Region-Wise)

Table 2.2.1: North India (no. of co-op societies)

State	80-81	90-91	2000-01	2010-11	2017-18	2018-19
Haryana	505	3229	3318	7019	7271	7264
Himachal Pradesh	0	210	288	740	959	977
Jammu & Kashmir	0	105	0	0	457	457
Punjab	490	5726	6823	7069	8018	7353
Rajasthan	1433	4976	5900	1290	14496	14882
Uttar Pradesh	248	7880	15648	21793	31133	31754
Uttarakhand	0	0	0	0	4112	4168
North Regional Total	**2676**	**22126**	**31977**	**52911**	**66446**	**66795**

Source: NDDB, 2020

Table 2.2.2: East India (no. of co-op societies)

State	80-81	90-91	2000-01	2010-11	2017-18	2018-19
Assam	0	117	125	155	374	402
Bihar	118	2060	3525	9425	21318	22261
Jharkhand	0	0	0	53	614	622
Meghalaya	0	0	0	0	97	97
Mizoram	0	0	0	0	39	42
Nagaland	0	21	74	49	52	52
Odisha	0	736	1412	3256	5857	5944
Sikkim	0	134	174	287	497	517
Tripura	0	73	84	84	101	100
West Bengal	584	1223	1719	3019	4066	4117
East Regional Total	**702**	**4364**	**7113**	**16328**	**33015**	**34154**

Source: NDDB, 2020

Table 2.2.3: South India (no. of co-op societies)

State	80-81	90-91	2000-01	2010-11	2017-18	2018-19
Andhra Pradesh	298	4766	4912	4971	3274	3308
Karnataka	1267	5621	8516	12372	15817	16021
Kerala	0	1016	2781	3666	3293	3317
Tamil Nadu	2384	6871	8369	10079	10806	10677
Telangana	0	0	0	0	244	5189
Puducherry	0	71	92	102	104	104
South Regional Total	**3949**	**18345**	**24670**	**31190**	**35735**	**38616**

Table 2.2.4: West India (no. of co-op societies)

State	80-81	90-91	2000-01	2010-11	2017-18	2018-19
Chhattisgarh	0	0	0	757	1082	1112
Goa	0	124	166	178	182	183
Gujarat	4798	10056	10679	14347	19044	19853
Madhya Pradesh	441	3865	4877	6216	9263	9151
Maharashtra	718	4535	16724	21199	20647	20652
West Regional Total	**5957**	**18580**	**32446**	**42697**	**50218**	**50951**

Source: NDDB, 2020

Region-Wise Dairy Co-Operative Infrastructure Capacity (Installed):

Table 2.2.5: North Region

State	BMC ('000 Litre)	Chilling Centre ('000 Litre. PD)	Dairy Plant ('000 Litre. PD)
Delhi	0	0	1500
Haryana	367	330	6775
Himachal Pradesh	136	80	100
Jammu & Kashmir	108	0	100
Punjab	1826	838	2185
Rajasthan	4375	510	3035
Uttar Pradesh	912	1580	3995
Uttarakhand	71	65	245
North Regional Total	**7795**	**3403**	**17935**

Source: NDDB, 2020

Table 2.2.6: East Region

State	BMC ('000 Litre)	Chilling Centre ('000 Litre. PD)	Dairy Plant ('000 Litre. PD)
Assam	20	0	60
Bihar	18332	314	2875
Jharkhand	186	10	690
Meghalaya	0	0	26
Mizoram	11	0	20
Nagaland	2	0	22
Odisha	785	80	680
Sikkim	9	0	60
Tripura	7	0	24
West Bengal	304	136	1267
East Regional Total	**3155**	**540**	**5724**

Source: NDDB, 2020

Table 2.2.7: West Region

State	BMC ('000 Litre)	Chilling Centre ('000 Litre. PD)	Dairy Plant ('000 Litre. PD)
Chhattisgarh	95	72	150
Goa	47	0	110
Gujarat	17530	6435	26295
Madhya Pradesh	854	599	1518
Maharashtra	2021	1855	11161
West Regional Total	**20547**	**8961**	**39234**

Source: NDDB, 2020

Table 2.2.8: South Region

State	BMC ('000 Litre)	Chilling Centre ('000 Litre. PD)	Dairy Plant ('000 Litre. PD)
Andhra Pradesh	1979	38	2705
Karnataka	4324	2960	8525
Kerala	1368	100	1910
Tamil Nadu	1594	1425	4121
Telangana	635	363	1250
Puducherry	50	0	120
South Regional Total	**9950**	**5286**	**18631**

Source: NDDB, 2020

Region-Wise Milk Procurement

Table 2.2.9: North Region Milk Procurement '000 kilograms per day

State	80-81	90-91	2000-01	2010-11	2017-18	2018-19
Haryana	33	94	276	511	562	441
Himachal Pradesh	0	14	24	60	60	70
Jammu & Kashmir	0	11	0	0	35	28
Punjab	75	394	912	1037	1766	1640
Rajasthan	138	364	887	1629	2845	2791
Uttar Pradesh	64	382	791	504	361	403
Uttarakhand	0	0	0	0	194	251
North Regional Total	**310**	**1259**	**2890**	**3741**	**5821**	**5624**

Table 2.2.10: East Region Milk Procurement '000 kilograms per day

State	80-81	90-91	2000-01	2010-11	2017-18	2018-19
Assam	0	4	3	5	30	33
Bihar	3	95	330	1091	1603	1891
Jharkhand	0	0	0	5	121	162
Meghalaya	0	0	0	0	12	13
Mizoram	0	0	0	0	6	7
Nagaland	0	1	3	2	3	4
Odisha	0	41	94	276	508	492
Sikkim	0	4	7	12	36	38
Tripura	0	3	1	2	6	6
West Bengal	31	52	204	273	188	226
East Regional Total	**34**	**200**	**642**	**1666**	**2512**	**2871**

Source: NDDB, 2020

Table 2.2.11: South Region Milk Procurement '000 kilograms per day

State	80-81	90-91	2000-01	2010-11	2017-18	2018-19
Andhra Pradesh	79	763	879	1371	1199	1229
Karnataka	261	917	1887	3742	7077	7475
Kerala	0	185	646	688	1260	1298
Tamil Nadu	301	1106	1618	2097	3039	3381
Telangana	0	0	0	0	657	737
Puducherry	0	26	45	35	54	55
South Regional Total	**641**	**2997**	**5075**	**7932**	**13287**	**14175**

Source: NDDB, 2020

Table 2.2.12: West Region Milk Procurement '000 kilograms per day

State	80-81	90-91	2000-01	2010-11	2017-18	2018-19
Chhattisgarh	0	0	0	25	79	104
Goa	0	16	32	38	64	64
Gujarat	1344	3102	4567	9158	21093	22920
Madhya Pradesh	68	256	319	588	1105	1012
Maharashtra	165	1872	2979	3053	3568	3998
West Regional Total	**1577**	**5246**	**7897**	**12862**	**25908**	**28098**

Source: NDDB, 2020

Region-Wise Liquid Milk Marketing

Table 2.2.13: North Region '000 Liters per day

State	80-81	90-91	2000-01	2010-11	2017-18	2018-19
Haryana	2	80	108	362	329	301
Himachal Pradesh	0	15	20	23	25	22
Jammu & Kashmir	0	9	0	0	27	25
Punjab	7	139	420	802	991	1005
Rajasthan	12	136	540	1505	2242	2253
Uttar Pradesh	1	326	436	380	886	990
Uttarakhand	697	1051	1524	3050	6380	6558
North Regional Total	**719**	**1756**	**3048**	**6122**	**11040**	**11311**

Source: NDDB, 2020

Table 2.2.14: East Region '000 Liters per day

State	80-81	90-91	2000-01	2010-11	2017-18	2018-19
Assam	0	10	7	22	51	54
Bihar	8	111	324	454	1126	1106
Jharkhand	0	0	0	253	383	391
Meghalaya	0	0	0	0	12	11
Mizoram	0	0	0	0	5	7
Nagaland	0	1	4	3	5	7
Odisha	0	65	98	290	409	403
Sikkim	0	5	7	17	35	41
Tripura	0	6	7	15	12	13
West Bengal	17	26	27	41	32	41
Kolkata	283	526	840	644	1141	1102
East Regional Total	**34**	**200**	**642**	**1666**	**2512**	**2871**

Source: NDDB, 2020

Table 2.2.15: West Region '000 Liters per day

State	80-81	90-91	2000-01	2010-11	2017-18	2018-19
Chhattisgarh	0	0	0	34	149	158
Goa	0	36	83	69	71	67
Gujarat	210	1052	1905	3237	5256	5433
Madhya Pradesh	39	279	244	495	876	856
Maharashtra	18	363	1178	2023	2783	2728
Mumbai	950	1057	1390	841	1952	1935
West Regional Total	**1577**	**5246**	**7897**	**12862**	**25908**	**28098**

Table 2.2.16: South Region '000 Liters per day

State	80-81	90-91	2000-01	2010-11	2017-18	2018-19
Andhra Pradesh	19	552	733	1565	1337	1308
Karnataka	166	889	1501	2661	3886	4010
Kerala	0	223	640	1092	1286	1296
Tamil Nadu	109	405	559	989	1038	1022
Telangana	0	0	0	0	803	830
Puducherry	0	22	43	93	98	97
Chennai	245	662	725	1025	1168	1227
South Regional Total	**641**	**2997**	**5075**	**7932**	**13287**	**14175**

Source: NDDB, 2020

2.3. Milk Producer Companies

The basic structure of milk producer's company works on the principles of patronage and they involve in business with members only. NDDB Dairy Services (NDS), the wholly-owned subsidiary of NDDB has facilitated the incorporation and operationalization of 15 Milk Producer Companies (MPCs). Out of the 15 MPCs, six were supported under National Dairy Plan-I (NDP-I), five are supported by the Tata Trusts/ Dairy Health and Nutrition Initiative (DHANI-I) and four are supported by the National/ State Rural Livelihoods Mission (NRLM/SRLM).

As of December 2019, 15 MPCs were operational in 73 districts of eight states with a membership base of 5.72 lakh (including 61 % smallholders with up to three milk animals) and covered 14,054 villages, and acquired an average volume of 28.41 lakh liters per day of milk during the year. Women members in the MPCs constitute about 53 per cent, a significant part of it is contributed by nine MPCs who have exclusively women membership. The members of these 15 companies contributed about Rs 141 crore towards share capital. They are expected to achieve a budgeted turnover of about Rs 5482 crore during the year 2019-20 (DAHD, 2019).

NDS continued to support the Six large Milk Producer Companies, namely Paayasin Rajasthan, Maahi in Gujarat, Shreeja in Andhra Pradesh, Baani in Punjab, Saahaj in Uttar Pradesh, and Bapudham in Bihar, who are also End Implementing Agencies (EIAs) under NDP I. In the successful implementation of sub-projects under NDP I, NDS provided the MPCs with project management support. Under Collaborative agreement with the Tata Trusts/ DHANII, NDS has facilitated setting up 5 MPCs namely, Sakhi in Alwar and Asha in Pali Rajasthan, Shwetdhara in Pratapgarh, Uttar Pradesh, Ruhaanii in Mansa, Punjab and Indujaa in Yavatmal, Maharashtra.

A total of 4 MPC projects are approved under National/State Rural Livelihood Mission (NRLM) to cover about 2300 villages and enrolled 1,07,000 members and achieve breakeven volume within 2-3 years of grant support. All National/ State Rural Livelihood Mission (NRLM) supported projects are green field projects mainly in districts with limited milk potential and market access. The MPCs are designed to have exclusive women membership comprising predominantly the members of SHG groups. Under the above arrangement, NDS facilitated setting up 4 MPCs namely, Muktaa in Sagar and Maalav in Rajgarh, Madhya Pradesh, Kaushikee in Saharsa, Bihar and Balinee in Jhansi, Uttar Pradesh. Balinee MPC operationalized during 2019-20, will cover 5 out of 7 districts in Bundelkhand region of Uttar Pradesh. Another MPC in Kota, Rajasthan is in the process of incorporation in the coming year (DAHD, 2019).

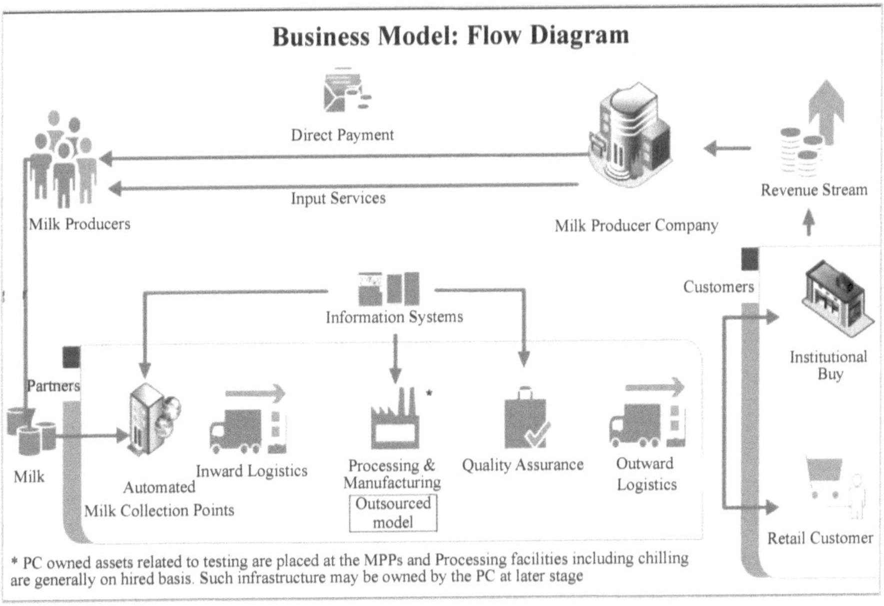

Fig. 2.2: Producer Company Business Model
Source: Singh S, 2015

2.4. Private Dairy Sector

Since 1991, when the era of industrial licensing reform began, private sector companies have experienced remarkable growth in capacity building for the production of milk and milk derivatives. They have made significant investments in the growth of capacity in the dairy sector, which in the past 20 years has exceeded the combined capacity of the dairy cooperatives and government dairies. Some of these private players are now much larger than some cooperative dairies and have significant growth potential. Since Private Sector functions purely on commercial lines to earn maximum profit, the social responsibility towards farmers' development is largely affected. The Private Players prefer to procure mil through vendors affecting the farmers' getting remunerative price. However, growth in Private Sector provides market access to a large number of farmers. As per FSSAI Licenses (until May 2019), the total number of Private Dairies (Milk Processing Units) is 1944 with a capacity of 901.6 LLPD (DAHD, 2019).

The coverage of organized private sector in the country is 10.2 per cent, while the organized Cooperative sector accounts for 10.3 per cent of total milk production. 'the rapid growth of private sector capacity in the last decade shows that the sector is prospering in the dairy sector, and its imprint is clearer in value-added products (FICCI, 2020).

It is estimated that the capacity created by private dairies in the last 20 years is more than the capacity set up by the cooperatives in over 30 years. Private dairy processing units are functioning in all the states; the highest concentration is in Uttar Pradesh and Maharashtra, accounting for over half of the total plants in the country. The total number of Private Dairies (Milk Processing Units) as per FSSAI Licenses (till May 2019) is 1944 with a capacity of 901.6 LLPD (DAHD, 2019). In terms of total installed capacity, the four states of Maharashtra, Uttar Pradesh, Gujarat, and Tamil Nadu account for more than half of the total (FICCI, 2020).

In the organized sector, private dairy companies are investing in developing an efficient milk procurement network as well as the marketing of liquid milk and valueadded products, in addition to focusing on bulk commodities like milk powder and ghee. Till the last decade, most of the private processors were operating on a small scale, focused on stabilizing their operations. Over the years, some of them have expanded to establish a strong regional base with aspirations to have a national presence in the next few years. It is noteworthy that the percentage of milk being handled by the private sector has exceeded that of the cooperative sector. This ratio is expected to further increase in favour of the private sector in the coming years (FICCI, 2020).

Some of the major organized private players include Tamilnadu-based Hatsun Agro Product Ltd which procures an average of 26 lakh liters per day (LLPD) or around one million tonnes (MT) annually. Besides, there are companies with an average milk procurement of 5-15 LLPD each. These include Parag Milk Foods, Schreiber Dynamix Dairies, Heritage Foods, Tirumala Milk Products, Sterling Agro Industries, VRS Foods, Nestle India, Prabhat Dairy, Indapur Dairy, Dodla Dairy, Creamline Dairy Products, SMC Foods,

Milkfood, Gopaljee Dairy Foods, and Anik Industries. (Major Status Quantity and products) State, company, year, procurement and products.

Table 2.3: Profile of Major Private Dairies

Company	Acquirer	Year	Location	Product
Vikram Dairy	Heritage Food India	2007	Maharashtra	Curd, Ghee, Paneer, Butter, Lassi, Ice Cream, Flavored Milk, Milkshake
Britannia New Zealand Foods	Britannia Industries	2009	Karnataka	Cheese, Butter, Dairy Whitener, Ghee, Flavored Milk, Paneer, Packaged Milk
Pashupati	Kwality	2010	Orissa	Packaged milk, milk, cultured products, ice-cream, frozen products, paneer, ghee and sterilized flavored milk.
Tirumala Milk Products	Carlyle	2010	Tamil Nadu	Cheese, Butter, Ghee, Plavoured Milk, Paneer, Packaged Milk
Wockhardt Limited	Danone	2011	Maharashtra	Milk Powder based nutritive products
Coco berry Retail Pvt Ltd	CX Partners	2011	Delhi	Frozen Yogurt
Parag Milk	IDFC PE	2012	Maharashtra, A.P.	Packaged milk, milk, cheese, yoghurt.
Prabhat Dairy	Rabo Equity	2012	Maharashtra	Ghee, Dahi, Ice cream, Paneer, Dairy Whitener, Cheese, Buttermilk, UHT Milk
Company	**Acquirer**	**Year**	**Location**	**Product**
Dodla Dairy	Cargill Ventures	2012	Gujarat	Butter, Ghee, Paneer, Curd, Flavoured Milk, Doodh Peda, Ice Cream and Skimmed Milk Powder
Cavin Kare	Chrys Capital	2013	Assam	Milk, Curd, Butter, Paneer, Milkshake, Buttermilk, Lassi
Shikhar Dairy	Samridhi Fund	2013	Delhi	butter, cream, ghee, cheese, khoya
Varshney Bandhu	Kwality Limited	2013	Delhi	butter, cream, ghee, cheese, khoya
Jyothi Dairy	Hatsun Agro	2014	A.P.	Packaged milk, milk,cheese, yoghurt.
Tirumala Dairy	Lactalis	2014	Tamil Nadu	Butter, Ghee, Paneer, Curd, Flavored Milk, and Skimmed Milk Powder
Westland Dairy	Milk Mantra	2014	Orissa	Probiotic Dahi, Pouch Dahi, Paneer, Lassi and buttermilk and Mishti Dahi.

Company	Acquirer	Year	Location	Product
Hatsun Agro	West Bridge Capital	2014	Chennai	Milk, Ice cream, Dahi, Paneer
Sahyog Clean Milk	Acumen Fund Advisory	2015	MP	Milk, Butter, Paneer, Ghee
Drum Foods International	Visha Narain Fireside	2015	Maharashtra	Curd, Misti Doi, Yogurt

Source: FICCI, 2020 The performance of the dairy industry will rely on the ability to develop a strong milk procurement network that will provide a cost advantage for raw materials along with regular supply assurance. Due to the inability to establish a solid network, most of the MNCs in this sector failed. Laboratory cases persisted in Moga (Nestle) and Etwah (Levers), though Anand (Amul) was replicated. Powerful backward integration in the form of cattle feed supply, vaccination, breed development projects, and farmers' profitsharing are some of the issues that could not be handled or exacerbated by any private sector firm. Primarily due to the strong involvement of cooperatives, the amount paid for milk in India is the highest in the world.

2.5. Milk Supply Chains

Milk marketing in India has largely remained an unorganised sector. Milk is marketed through the fragmented supply chains inclusive of the local milk vendors, wholesalers, retailers. The existing structure of milk marketing is an outcome of India's demographics, traditions, infrastructural developments, socioeconomic empowerment of milk producers, and institutional initiatives. Several types of milk-marketing chains presently operate in India. The dairies under the public sector, cooperatives, and organised private multinationals constitute the formal milk-marketing chains, and the remaining constitute the traditional or informal milk-marketing chains (Figure 2.3). Though there may not be uniformity in the estimates of marketed surplus milk, there is a consensus on the predominance of the informal sector in the Indian milk market.

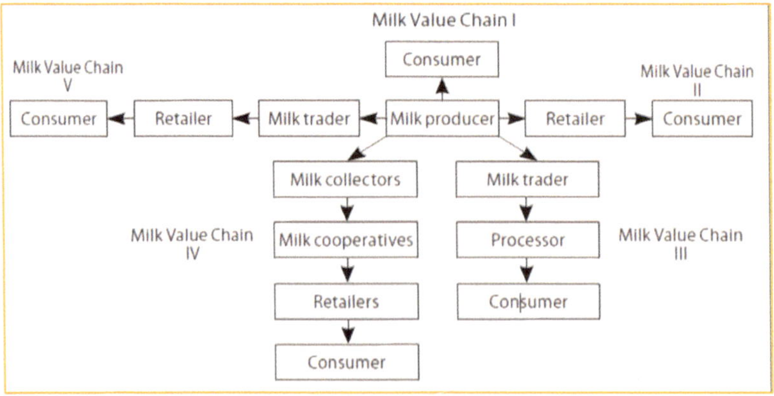

Fig. 2.3: Different Milk Supply Channels in India
Source: Kumar *et al*, 2018

2.6. NCDFI – E Marketing

Majority of Indian population is engaged in agriculture and allied sector activities for their livelihood. Bringing efficiencies in the marketing of dairy and agricultural commodities ensures maximum and measurable benefits for the entire agriculture and allied ecosystem in the country. Utilizing technology to create robust value chains is the way forward in achieving this goal. With this objective, National Cooperative Dairy Federation of India (NCDFI), with the technical support of NCDEX e-Markets Limited (NeML) has launched NCDFI e-Market on June 10, 2015. It is an electronic marketplace for efficient buying and selling of bulk commodities across the Country. The commodities being traded on NCDFI eMarket platform are Dairy Products, Cattle Feed Ingredients, Edible Oils, Packaging Materials, Sugar, Chemicals, Scrap, etc. apart from Transport Services.

NCDFI eMarket offers

1. Efficient Price Discovery Mechanism
2. Quality Produce
3. Fast and User-friendly Platform
4. Reduced Transaction Costs
5. Increased Transparency & Auditability

NeML offers e-Pledge model in which NeML act as facilitator, bringing borrower, Warehouse Service Provider (WSP) and Bank under one roof, providing warehouse and commodity electronic accounting system, Pledging, Trading, clearing and settlement solution. Borrowers (farmers / Traders / processors / FPOs etc.) after depositing commodities in NeML notified warehouses can apply on-line to Bank via Comlive and avail finance against pledge of warehouse receipts.

The physical warehouse receipt financing has been in existence for a long period of time where banks lend against the physical warehouse receipts issued by warehouseman. Due to lack of MIS, redundant risk management process, and lack of transparency Banks have faced many frauds and scams. NeML's e-pledge warehouse-based commodity finance model work towards removing such malfunctioning of the system by offering better risk management, enhanced operations and technological solutions. NeML's empanelled Warehouse Service Provider (WSP) act as Collateral management agency responsible for commodity safety, security, stock management, insurance, deposit and delivery whereas NeML provides its e-pledging platform and better risk management process. E-pledge model ensures safe, speedy, smooth and transparent commodity finance business transaction.

NeML also offers e-auction of defaulted stock (even in physical form) on its trading platform which has large pool of buyers already registered at pan India level and Banks realize better prices for such commodities.

Benefits

1. E-pledge is simple, secure and speedy mode of commodity accounting and pledging platform.
2. Comlive Platform brings more visibility, traceability in the existing practices of WHR finance and make entire process more secure.
3. Enhanced monitoring of e-pledge transaction through regular e-mail alert services for various aspects like pledge creation, Pledge redeem, price drop, insurance expiry, revalidation of commodities etc.
4. COMLIVE Platform ensures pledge monitoring and real time data accountability via providing multiple customizable MIS reports.
5. Regular internal as well independent third-party audit and inspection of warehouses and commodities to Maintain sanity of the operations
6. Facility of online trading of pledged commodity, which leads to better realization to the borrowers.
7. In case of Credit default, Default auction on NeML Platform for disposal and better Price Discovery

Auction Process

1. **Forward Auction:** Forward Auctions are used by NCDFI registered sellers to sell their products to potential buyers. This is an ascending price forward auction where price is increased round by round only if demand exceeds supply. Under this auction, the price announced is at the discretion of the seller. Bidders bid the volume they wish to purchase at the declared price. As multiple sellers participate parallel in these auctions, quoted prices remain competitive. Post auctions, the remaining quantities, if any, would be made available throughout the week at 'Continuous Markets'.
2. **Reverse Auction:** Reverse auction mechanism allow buyers to seek competitive prices by inviting sellers to participate in real time auction. The required product, quantity, specifications, place of delivery and delivery period are announced by the buyer prior to the auction date.

Interested sellers can participate in the auctions by bidding an all-inclusive FOR prices. Lowest quoted price is conveyed to the buyer at the end of the auction for approval. Upon acceptance by the buyer, contract notes are issued to the successful bidder.

3. **Auction Sale:** This is the product developed to replace the traditional tenders and further negotiations. This is a Yankee Forward Auction where the sellers offer the products available with them while buyers bid both prices and their required quantities. Multiple prices can be discovered at the end of the auction and if the bid prices are acceptable to the Seller, then the contracts would be concluded.

4. **Auction Purchase:** This is the product developed to replace the traditional tenders and subsequent negotiations for procurement of agri-commodities. This is a Yankee Reverse Auction where the buyers put up auction for their required products while sellers bid for their intended quantity and price, they are willing to offer in auction. On real time basis, system will display the tentative quantities allocated to each bidder. Multiple prices can be discovered at the end of the auction and if the bid prices are acceptable to the Buyer, then the contracts would be concluded."

2.7. Gujarat State - Dairy Industry Profile

Gujarat State is a leading State for milk production in the country. "Amul" pattern is well known and accepted by all the states in our country and some of the other countries also. Milk procurement by co-operative movement is basic theme in our state. At village level 18932 Milk Co-operative Societies, 102 Chilling centres and 18 Dairy processing units at 22 district level (Dairy) are in functioning in the state. Reaming four dairy union of Kachchh-Saurastra region is assisted for establishment of processing plant under RKVY scheme. They collected total of 86516.193 Lakh Kg milk and processed. State government has given full support to Dairy Development through Dairy Co-operation movement.

22 Co-operative Dairy Unions have total 232.45 LLPD milk processing capacity and they produced 237.03 LKPD milk in the year 2018-19. These 22 dairy unions have 102 chilling centres also having capacity of 72.42 LLPD milk (DoAH, GoG, 2020).

Table 2.4: Profile of Gujarat State Dairy Co-operatives

Sr.No	Dairy Cooperative	No. of Cooperative Dairies	Milk Processing Capacity LLPD	Products Range
1	Ahmedabad	680	10.5	NA
2	Amreli	790	2	Ghee, Buttermilk, Dahi
3	Banaskantha	1499	56	SMP, Ghee, Cheese, Dairy Whitener, Paneer, Dahi, Ice Cream, Butter, Whey Powder, UHT Milk
4	Bharuch	754	3	Ghee, Cream, Khoa, Buttermilk, Paneer, Flavoured Milk, Dahi, White Butter, Masala Chhash
5	Bhavnagar	677	5	Ghee, Buttermilk, Dahi, Paneer,
6	Botad	255	0	Dahi, Paneer
7	Dev-Bhumi Dwarka	226	0	Chocolate, Dairy Whitener, Paneer, Flavored Milk, Dahi, Margarine, White Butter
8	Gandhinagar	124	0	Ghee, Cream, Buttermilk, Ice cream, Sweets, Dairy Whitener, Dahi, Table Butter, White Butter
9	Jamnagar	142	0.45	NA
10	Junagadh	362	2	Ghee, Buttermilk, Dairy Whitener, Flavored Milk, Dahi, Table Butter
11	Kachchh	799	2	Dahi, White Butter
12	Kheda	1240	26	Milk, SMP, WMP, Butter, Ghee, Cheese, Buttermilk, Dairy Whitener, Paneer, Flavored Milk, Table Butter, White Butter, UHT Milk
13	Mehsana	1543	26	SMP, WMP, Ghee, Dairy Whitener, Dahi,
14	Morbi	182	0	Ice Cream, Dahi
15	Panchmahals	2162	18.5	SMP, Ghee, Buttermilk, Paneer, Flavored Milk, Dahi, Table Butter, White Butter, Jeera Chhash
16	Porabandar	869	2	Shrikhand
17	Rajkot	836	6	Ghee, Cream, Buttermilk.
18	Sabarkantha	1922	40	SMP, Ghee, Khoa, Butter, Dairy Whitener,
19	Surat	1184	15	Shrikhand, Dahi
20	Surendranagar	766	2	Butter
21	Vadodara	1454	8	Lassi, Gulabjamun, Butter, Ghee, Cream, Khoa, Cheese spread, Buttermilk, Penda,
22	Valsad	1112	8	Ice Cream, Sweets, Shrikhand,

Source: DAHD, 2019

Table 2.5: Profile of Private Dairies in Gujarat

Sr. No	Dairy	Location	Products
1	Baba Casein International	Vadodara	Butter, Ghee, Cream, Casein, Mayonnaise, Margarine
2	Clarion Casein Ltd.	Kadi	Industrial Casein, Rennet Casein, Sodium Caseinate, Calcium Caseinate, Edible Acid Casein
3	Classik Casein Ltd.	Anand	Sodium Caseinate, Calcium Caseinate, Technical Casein, Industrial Casein, Edible Casein
4	Doshi Brothers	Anand	Casein - industrial grade, technical grade, edible grade, different grades of acid casein
5	Dudhia Dairy Products	Ahmedabad	Dairy Sweets, Paneer, Ghee
6	Fable Food Products Pvt. Ltd.	Ahmedabad	Ghee
7	Gagar Milk Foods	Kadi	Fat Spread, Margarine, Mayonnaise
8	Gamdiwala Dairy	Ahmedabad	Milk, Butter, Ghee
9	Gangotri Dairy Products Pvt. Ltd.	Porbandar	Milk, Buttermilk
10	Gayatri Dairy Products Pvt. Ltd.	Mehsana	Milk, Ghee, Skimmed Milk Powder / Whole Milk Powder / Dairy Whitener.
11	Lactose India Ltd.	Vadodara	Lactose Monohydrate
12	National Casein Co	Nadiad	Industrial Casein, Rennet Casein, Edible Acid Casein
13	P S Milk processors Pvt. Ltd.	Ahmedabad	Milk, Ghee
14	Rich Milk (India) Pvt. Ltd.	Ahmedabad	Acid Casein, Buttermilk, Processed Cheese, Curd, Ghee,
15	Royal Dairy Ltd.	Bharkawada	Dairy Sweets, Ice Cream, Lassi
16	Samruddh Dairy Products (P) Ltd.	Ahmedabad	Milk, Butter, Ghee
17	Shah Milk Processors Pvt. Ltd.	Rajkot	Milk, Butter, Ghee
18	Sundhiawala Dairy	Ahmedabad	Milk, Ghee
19	Vadilal Industries Ltd.	Pundra	Ice Cream
20	Vimal Dairy Ltd.	Mehsana	Milk, Cheese, Ghee, Butter, and Skimmed Milk Powder.

Source: Dairy India, 2018

2.8. Unorganised Dairy Sector

The Indian dairy industry is divided into the organized and unorganized segments. The unorganized segment consists of traditional milkmen, vendors and self-consumption at home, and the organized segment consists of cooperatives and private dairies. Unorganized sectors have predominance over the Indian Dairy Industry. On account of the same, despite India being the largest producer of milk and milk products, none of the major Indian dairy companies feature on the list of top 15 major dairy giants, while one makes it to the list of top 20 dairy companies in the world. As per the Annual Report for FY19 of Dept. of Animal Husbandry, Dairying & Fisheries, Ministry of Agriculture & Farmers Welfare, GOI, co-operatives & private dairies still procure only about 20 per cent of the milk produced in the country, while 32 per cent is sold in the unorganized market and about 48 per cent is consumed locally (Sabnavis *et al*, 2020).

The dairy product industry has relatively low product differentiation. At the same time the industry is unorganized with a large number of small players. Organized dairy industry accounts for around 20 per cent of the total milk produced. Approximately half of the milk produced is marketed. Out of the total milk distributed jointly by the organized and unorganized segments, about 50 per cent is consumed in milk form and the balance is converted into various milk products like milk powder, ghee, butter, cheese, yoghurt, etc. Cost effective technologies, mechanization and quality control measures are seldom exercised in the unorganized sector and remain key issues to be addressed (Bhagyalaxmi, 2020).

2.9. Milk Adulteration

Normally, the adulteration in food is done either for financial gain or lack of proper hygienic conditions of processing, storing, transportation and marketing. This ultimately leads to the stage that the consumer is either cheated or becomes victim of diseases. Such types of adulteration are quite common in developing countries.

Milk is an important source of nutrient required for growth in infants and children and for maintenance of health in adults. Milk is a perfect food, readily digested and absorbed. It is a sole natural food for infants and children. But these days it is being adulterated with harmful substances which enhance its quantity and characteristics but reduces its quality. Adulterants are mainly added to increase the shelf life of milk. Adulteration in milk is considered to reduce the quality and to increase the quantity of milk.

National Milk Safety and Quality Survey Report 2018 by FSSAI, which was the largest sample survey done so far, revealed that 48.2 per cent of the samples were found to be noncompliant (DAHD, 2019). While the survey results effectively counter wrong perception of large-scale milk adulteration in India, but non-compliance on quality parameters, particularly in processed milk is a matter of concern. The survey has shown that about 41 per cent samples, though safe, fall short of one or another quality parameter or standard. There is non-compliance on account of low fat or low SNF (solids not fat), two key quality parameters both in raw and processed milk. In raw milk, proportion of fat and solids not fat (SNF) varies widely by species and depends on breed as well as quality of feed and fodder. Cattle must be properly fed and good farm practices must be adopted to improve the amount of fat and SNF in milk. Thus, low fat and SNF for these reasons or due to dilution of milk with water in understood (FSSAI, 2018).

Milk is produced throughout the year. However, milk production is greatly reduced during summer months due to heat stress and scarcity of fodder. Milk is transported from point of production to cities mainly through middlemen. Such milk is watered/ skimmed to increase profit. To maintain its composition, materials like starch, flour, urea, cane sugar and vegetable oil are added as adulterants. Milk dealers may either dilute the milk or extract valuable components and there after add cheap substances to maintain its compositional parameters. These substances include starch, urea, preservatives like formalin, hydrogen peroxide, boric acid and various antibiotics. Urea is added for whitening of milk and only few grams of urea are enough to bring milk in its original state. Hydrogen peroxide is used as preservative usually in summer season when environmental temperature is very high. This unethical activity is usually adapted to prevent the financial losses due to the spoilage of milk during its transportation and sale (Brindha *et al*, 2017)

3

Trade Performance of Indian Dairy Sector

Cheese, butter, whole milk powder, and skimmed milk powder are a major part of the world's dairy trade. The demand for worldwide dairy exports is highly concentrated. With a share of 12 per cent, New Zealand is the largest exporter of dairy products, followed by Germany, Netherlands, France and the US. Germany has been the world's major dairy importer as well with a share of 10 per cent. Other large dairy product importers are China, Netherlands, France, Italy and Belgium.

Before 1999, India was a net importer of dairy products. The successful implementation of the 'Operation Flood' program and trade reforms caused the direction of India's dairy trade to change marginally. India has been a net exporter of dairy products since 2000. Statistics show that India's export share of dairy products over world dairy exports increased marginally from 0.1 per cent in 2001 to 0.5 per cent in 2019, while its import share of dairy products over world dairy imports increased marginally from 0.018 per cent in 2001 to 0.04 per cent in 2019. The Middle East, South Asia, South East Asia, and the USA are the major markets for India's dairy products. Conversely, India is a favorable market for the European Union, New Zealand, Australia, UAE, and US dairy products.

3.1. Country Wise Trends of Export in Dairy Sector

Theoretically speaking, the value of exports of a particular commodity from one country should match the value of its imports in another country for the year under consideration. In practice, however, one often encounters some discrepancy because the exports are recorded FOB (Free on Board) while imports are recorded under CIF (Cost, Insurance and Freight) (Joshi, 2012). Therefore, we have provided India's dairy exports and imports – to and from the world – separately. In this section, the country-wise exports of dairy products as a whole, from 2001 to 2019, have been presented.

India's export share of milk products increased slightly, from 0.1 per cent in 2001 to 0.5 per cent in 2019, while its share of milk production increased from

14 per cent to 22 per cent during the same period. Overall, the data shows that the top 15 countries, also major players in dairy exports, are all developed nations.

Table 3.1: Country-Wise Status Dairy Export

Country	2001-02	2008 - 09	2016 - 17	2018 - 19	Rank Based on 2018 - 19	CAGR (%) (2001-02 to 2018 – 19)
	Value 000 USD	Value 000 USD	Value 000 USD	Value 000 USD		
New Zealand	2664188	6562092	7819257	10645288	1	8.00
Germany	4324516	9478290	8089599	10250756	2	4.91
Netherlands	3331185	6907739	7098551	9609572	3	6.06
France	3500867	7379107	6301599	7075381	4	3.99
USA	598650	2977259	3328278	4968490	5	12.48
Belgium	1743558	3397511	3004900	4174646	6	4.97
Italy	945503	2309230	2975661	4074109	7	8.45
Ireland	947217	2096019	1912793	3386995	8	7.34
Poland	369548	1717809	1589521	2769627	9	11.84
Denmark	1286544	2562313	2280439	2715799	10	4.24
UK	781924	1497435	1632948	2424353	11	6.49
Belarus	148656	1090112	1815082	2255487	12	16.31
Australia	1555839	2207709	1628380	1794950	13	0.80
Spain	499591	1377478	1221282	1654604	14	6.88
Austria	507442	1403528	1289440	1465895	15	6.07
India	39714	269995	130531	460406	30	14.58
Top 15	22827138	52223554	51853852	69265952		
Rest of the world - 215	4564858	13920034	12798746	19686470		
Total	27391996	66143588	64652597	88952422		

Source: ITC Trade map, 2020

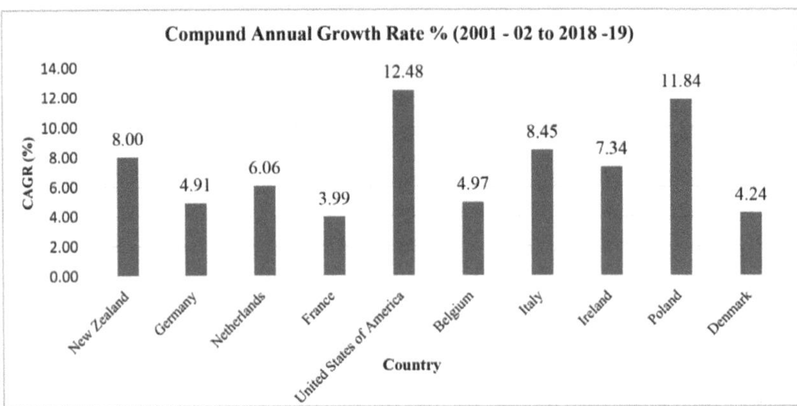

Fig. 3.1: Country Wise Growth Rate of Dairy Export

Source: ITC Trade map, 2020

Table 3.1 shows the dairy exports of major 15 countries in the world. In 2001, Germany exported around 16 per cent of its dairy products to the world, which was the highest; this figure slightly declined to almost 12 per cent in 2019. Similarly, New Zealand's export share of milk products increased slightly, from 10 per cent in 2001 to 12 per cent in 2019 being the largest dairy product exporter. The Germany's exports accounted for around 12 per cent, followed by Netherlands (10.80 %), France (7.95 %) and the US (5.59 %), and so on in 2019. Table 3.1 also reveals that more than 50 per cent of dairy products were exported by the top five developed countries and 80 per cent by the top 15 countries. It may also be noted that India is a minor player in the world market in terms of exporting dairy products while being ranked the world's single largest country producing dairy animals (including dairy cows and buffalo).

India's share of exports of dairy products increased slightly from 0.1 per cent in 2001 to 0.5 per cent in 2019, while its share of milk production increased from 14 per cent to 22 per cent in the same period. Overall, the data shows that the top 15 countries, also major players in dairy exports, are all developed nations. Milk production, on the other hand, is mainly dominated by five countries namely India, the USA, Pakistan, China, and Brazil (Parida *et al*, 2019).

3.2. Country Wise Trends of Import in the Dairy Sector

Table 3.2 shows the country-wise import of dairy products covering the period between 2001 and 2019. It shows that Germany was the leading importer of dairy products, dominating 9.59 per cent of the world's total dairy in 2019. Next in line is China (7.05 %) followed by Netherlands (5.19 %), the France (4.98 %), Italy (4.80 %), and the Belgium (4.53 %). Except for China & Saudi Arab (which is a developing nation) 61 per cent of dairy products were imported by the top 15 developed countries in that period. The data shows that the import share of dairy products of most top 15 countries declined over the period 2001-2019, barring a few nations including China, Russian Federation, Hong Kong, Saudi Arabia, and the UAE where it increased. Interestingly, China's demand for milk increased manifold over the same period. While China's dairy industry continues to grow rapidly, its consumption has outpaced domestic supply. It is for this reason that exporters from New Zealand, Europe, Australia, and the United States are locating new opportunities in China's dairy market (Parida *et al*, 2019).

Table 3.2: Country-Wise Status Dairy Import

Country	2001-02	2008 - 09	2016 - 17	2018 - 19	Rank Based on 2018 - 19	CAGR (%) (2001-02 to 2018 – 19)
	Value 000 USD	Value 000 USD	Value 000 USD	Value 000 USD		
Germany	2844278	6780742	5990836	8638110	1	6.37
China	215942	861671	3371298	6352140	2	20.67
Netherlands	1645361	3189480	3345020	4674823	3	5.97
France	1844363	3318301	3241572	4483812	4	5.06
Italy	2397728	4777436	3487660	4319450	5	3.32
Belgium	1862070	3431689	3134099	4080048	6	4.45
UK	1666528	3833549	3179704	3948360	7	4.91
Russian Federation	391343	1485723	1906597	2966997	8	11.91
USA	1021305	1679549	1966013	2788670	9	5.74
Hong Kong, China	356389	502564	1807866	2312642	10	10.95
Spain	1029530	2627416	1623210	2170083	11	4.23
Mexico	787863	1469842	1471133	2095609	12	5.59
Saudi Arabia	539947	1636858	1784742	1921938	13	7.31
Japan	740668	1339277	1204863	1920424	14	5.44
UAE	271844	814572	1296716	1726308	15	10.82
India	4861	15176	40436	33849	123	11.38
Top 15	17615160	37748669	38811329	54399414		
Rest of the world	9881845	26109437	25343325	35673497		
Total (205)	27497005	63858106	64154653	90072911		

Source: ITC Trade map, 2020

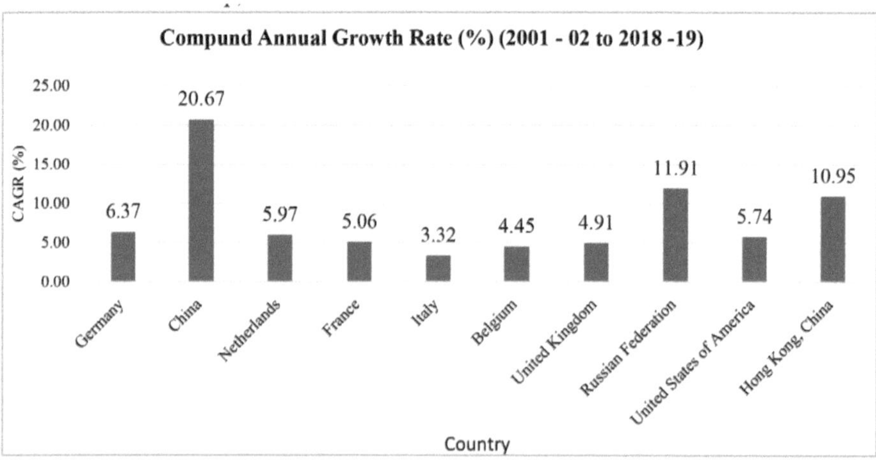

Fig. 3.2: Country Wise Growth Rate Dairy Imports
Source: ITC Trade map, 2020

As far as India's trade is concerned, it imported 0.018 per cent of its total world dairy imports in 2001; this marginally increased to 0.04 per cent in 2019. In a nutshell, the data shows that India remains a minor player in terms of imports, holding the 97th position, along with exports where it is ranked 30th in the world for trade in dairy products.

3.3. International Market for India's Dairy Products

Table 3.3 shows the major export destinations for India's dairy products. India's dairy exports increased from US$ 97.03 million in 2006 to US$ 460.41 million in 2019; exports in the major 15 countries showed the same trend except for Bangladesh. South Asia, the Middle East, South East Asia, and the US are the major markets for India's dairy products. India's export share of dairy products to the top 15 countries comprised 61 per cent in 2006 and reached 72 per cent in 2019. Out of this 72 per cent, India's export to the Middle East (32 %) ranked highest followed by the US (20 %), South Asia (11 %) and East Asia (6.8 %) in 2019. Rising demands for Indian dairy products in the Middle East, South Asia, and Southeast Asia may be attributable to lower transportation costs, taste, and cultural proximity.

Table 3.3: Country-wise distribution of India's export of the dairy products

Exports (Values in Million US$)					
Country	**2006 – 07**	**2010 - 11**	**2014 - 15**	**2017 - 18**	**2018 -19**
USA	2.26	2.95	3.51	9.78	89.5
UAE	14.55	23.71	35.66	38.45	42.1
Oman	1.51	3.27	6.29	8.02	38.4
Saudi Arabia	2.61	3.5	4.58	5.62	25.7
Egypt	12.43	14.68	0.21	26.85	23.6
Bhutan	0.33	0.92	7.31	19.7	21.7
Indonesia	2	5	7	5	14.5
Bangladesh	10.8	14.51	37.35	10.2	14.1
Singapore	2.34	3.45	6.49	10.43	11.3
Morocco	4	4	5	2	10.3
Nepal	4.94	9.59	13.66	12.5	9.7
Qatar	0.18	2.01	3.77	8.16	9.6
Maldives	0.435	3	5	6	9.0
Kuwait	0.87	2.61	4.63	5.58	6.7
Rest of the World - 215	35.5	29.15	36.85	17.84	64.27
Total	97.03	126.22	207.45	209.87	460.41

Source: ITC Trade map, 2020

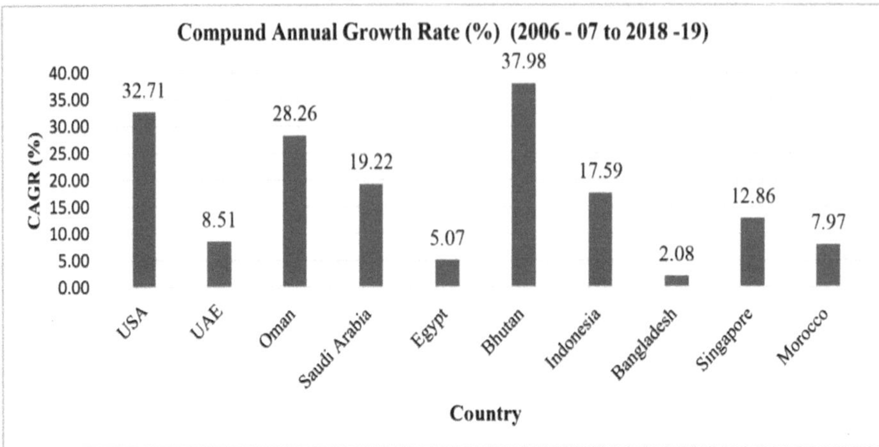

Fig. 3.3: Country-wise Growth Rate of India's Export of the Dairy Products
Source: ITC Trade map, 2020

India's dairy exports to its neighbours, Bangladesh and Pakistan, had declined between 2006 and 2019. This could be attributed to the prevailing geopolitical factors coupled with improvements in domestic production. Notably, Indian dairy products have been unable to penetrate European and South American markets.

Table 3.4 shows the country-wise distribution of India's imports of dairy products over the period 2006-2019. India's dairy imports increased from US$ 23.5 million in 2006 to US$ 33.85 million in 2019. In 2019, India imported more than 33 per cent of dairy products from France followed by Germany (9.80 %), Italy (9.66 %), Denmark (5.91 %), United Kingdom (5.47 %), and so on. It may be observed that almost 71 per cent of dairy products imported by India were from the European countries. Overall, the data shows that the majority of India's dairy imports were mainly from developed nations. There were a few exceptions like Nepal with a share of around 1.33 per cent in terms of total dairy imports by India in 2019.

Table 3.4: Country-wise distribution of India's Imports of the dairy products

Imports (Values in Million US$)					
Country	**2006**	**2010**	**2014**	**2017**	**2019**
France	2.7	4.3	11.9	11.3	11.49
Germany	0.3	0.3	2.2	2.9	3.32
Italy	0.5	1.5	2.6	2.9	3.27
Denmark	1.9	3.4	2.7	3.4	2
UK	0.4	2.8	2.6	1.9	1.85
Turkey	0.2	0.004	2.93	4.68	1.52

Imports (Values in Million US$)					
Country	**2006**	**2010**	**2014**	**2017**	**2019**
New Zealand	12.5	113.2	6.3	2.1	1.37
Spain	0.13	0.19	0.42	2	1.1
Netherlands	0.8	2.3	4.2	1.7	1
USA	0.7	2.4	4.1	0	0.56
Nepal	2.1	2	1.6	1.4	0.45
Rest of World	1.3	2.5	7.9	15.4	5.81
Total	23.5	180.2	46.3	43.8	33.85

Source: ITC Trade map, 2020

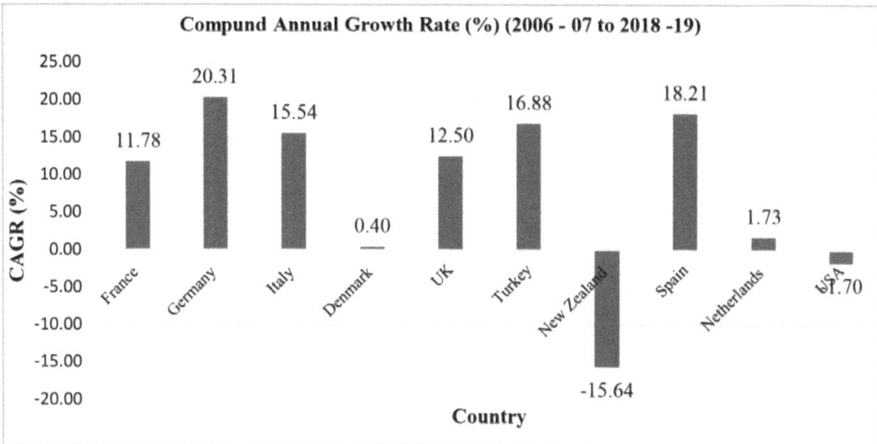

Fig. 3.4: Country-wise Growth Rate of India's Import of the Dairy Products
Source: ITC Trade map, 2020

Table 3.5 shows a product-wise (HS 4 digit) distribution of India's dairy exports over the period 2006-2019. India's export share of milk and cream, not concentrated products, remained stagnant, from 2.18 per cent in 2006 to 2.07 per cent in 2019 with a CAGR of -0.39 per cent. Similarly, the export share of butter and other fats and oils, along with cheese and curd products, increased from 14.11 per cent to 43.70 per cent and 2.20 per cent to 7.51 per cent, respectively, in the same period with a CAGR of 9.08 per cent and 9.91 per cent, respectively. The export share of milk and cream concentrated (dairy products containing added sugar or other sweeteners) (HS 0402), buttermilk, curdled milk and cream (e.g., fermented or acidified milk and cream which may contain added sugar or other sweeteners, flavouring agents, fruits, nuts, or cocoa) (HS 0403), and whey and other natural milk constituents (HS 0404) products showed a declining trend. Their CAGR evinced a negative growth in the same period.

Table 3.5: Product-wise export share of the dairy products in India

HS Codes	Dairy Products	2006 - 07	2010 - 11	2014-15	2017-18	2019 - 20	CAGR %[#]	Major Destinations
401	Milk and cream, not concentrated	2.18	1.96	4.96	3.34	2.07	-0.39	Bhutan, Singapore, Nepal, UAE, China (Hong Kong)
402	Milk and cream, concentrated	91.10	48.21	113.24	19.76	7.49	-17.48	Bangladesh, Bhutan, Afghanistan, Nepal, Pakistan
403	Buttermilk, curdled milk and cream	0.90	0.394	0.22	0.70	0.25	-9.38	Singapore, USA, Bhutan, UAE, New Zealand
404	Whey and other natural milk constituents	2.00	1.88	0.07	0.21	0.16	-17.58	UAE, Nepal, Singapore, Oman, USA
405	Butter and other fats and oils	14.11	32.05	22.36	38.76	43.70	9.08	Turkey, UAE, Egypt, Saudi Arabia, USA
406	Cheese and curd	2.20	6.88	9.34	14.12	7.51	9.91	UAE, Bhutan, USA, Singapore, Saudi Arabia

CAGR calculated here is from the year 2006 - 07 to 2019 - 20

Source: ITC Trade map, 2020

From the above table 3.5, it is easily discernible that more than 40 per cent of Indian exports hailed from the butter and other fats and oils (e.g., butter, butter oil/ghee, dry spreads) (HS 0405) category in 2019. This was followed by cheese and curd (HS 0406) (7.51 %), milk and cream, concentrated (HS 0402) (7.49 %), and milk and cream not concentrated (HS 0401) (2.07 %); the latter's export share was the lowest among all the dairy products. In sum, the Study concludes that there is greater scope for enhancing the export of some products including milk and cream, not concentrated, butter and other fats and oils and, cheese and curd.

Table 3.6: Product-wise import share of the dairy sector in India

HS Codes	Dairy Products	2006 - 07	2010-11	2014-15	2017-18	2019 - 20	CAGR %#	Major Destinations
401	Milk and cream, not concentrated	0.17	0.09	1.45	2.02	3.52	25.99	France, Italy, Ireland, Denmark, Belgium
402	Milk and cream, concentrated	6.13	50.51	14.00	9.77	5.72	-0.53	France, New Zealand, Saudi Arabia, Netherlands, Belgium
403	Buttermilk, curdled milk and cream	0.79	0.31	0.51	3.35	3.75	12.72	Bangladesh, Australia, Switzerland, United Kingdom, Bhutan
404	Whey and other natural milk constituents	13.91	6.29	60.31	48.24	43.02	9.07	France, Germany, Turkey, Norway, Argentina, UK
405	Butter and other fats and oils	58.51	41.01	6.77	12.03	6.25	-15.81	New Zealand, Nepal, Turkey, France, Belgium
406	Cheese and curd	12.71	3.84	18.72	18.87	29.88	6.80	Italy, Denmark, UK, Netherlands, Germany

CAGR calculated here is from the year 2006 - 07 to 2019 - 20

Source: ITC Trade map, 2020

Table 3.6 shows the product-wise import distribution of four-digit HS dairy products over the period 2006-2019 and its CAGR. In 2019, India imported more than 40 per cent whey products followed by cheese and curd (HS 0406) (29.88 %), butter and other fats (HS 0405) (6.25 %), milk and cream, concentrated (HS 0402) (5.72 %), buttermilk, curdled milk and cream (HS 0403) (3.75 %) and milk and cream, and not concentrated (HS 0401 -dairy products without added sugar or other sweeteners) (3.52 %). The latter registered the loweSt import share out of the product groups. Overall, the data indicates that the import share of all products increased over the period 2006-2019, except for butter and other fats and oils & milk & cream - concentrated.

3.4. Export Trends for Dairy Products of India

In India, milk production is growing by 6.4 per cent during the last five years and has increased from 146.3 million MT in 2014-15 to 187.7 million MT in 2018-19. However, India exports a relatively smaller volume of dairy products and has a small share in the global dairy trade despite being the largest milk producer in the world. Dairy products exported from India are mainly Skimmed Milk Powder (SMP), butter, butter oil, cheese, ghee, and buttermilk. SMP has a share of around 30 per cent in the exports of dairy products from the country. The major export destinations for dairy products are Turkey, United Arab Emirates, Egypt, Bangladesh, Bhutan, the United States, Saudi Arabia, and Malaysia.

Table 3.7: India's Dairy Product Exports (2014 – 15 to 2019 – 20)

	Quantity (tons)	Value (US Million $)
2014 - 15	66424	198
2015 - 16	33442	115
2016 – 17	39166	135
2017 – 18	48039	185
2018 – 19	180688	483
2019 - 20	111146	280
CAGR (%)		
(From 2014 – 15 to 2019 – 20)	**8.96**	**5.95**

Source: APEDA, 2020 (Quantities and Values includes skimmed milk powder, butter, butter oil, cheese, ghee, buttermilk).

Major Indian Dairy Products Export

1. Skimmed Milk Powder (04021010): Skimmed milk powder is obtained by the process of spray-drying fresh skimmed milk. The product obtained by the removal of fat and subsequent pasteurization is called skimmed milk. The raw material is fresh bovine milk i.e., the mixture of cow and buffalo milk to which no preservatives, neutralizers, or any other chemicals are added. It is mostly used in the preparation of Reconstituted milk, yogurt, sweetened condensed milk, ice cream mixes, and confectionery products.

Global leaders in the export of these products are the USA, Germany, New Zealand, France, and Belgium. Indian export of skim milk powder represents 1.2 per cent of global export ranking 15th. The major export destination for Indian skim milk powder is Bangladesh, Afghanistan, Malaysia, Pakistan, and UAE.

Table 3.8: Skim Milk Powder Export time-series data

Skim Milk Powder (04021010)		
	Quantity (MT)	**Value (Rs. Crore)**
2007 - 08	32050.05	450.71
2008 - 09	24360.5	376.55
2009 - 10	12173.5	137.05
2010 - 11	11344.59	154.67
2011 - 12	10.26	0.05
2012 - 13	69633.56	1085.55
2013 - 14	124017.8	2605.12
2014 - 15	27553.92	613.98
2015 -16	13838.21	251.44
2016 - 17	14698.17	266.33
2017 - 18	11307.88	216.11
2018 - 19	44409.5	549.54
CAGR %	**8.47**	**12.38**

Source: APEDA, 2020

2. **Butter (04051000):** In the year 2018, the export worth of 8,126,313 thousand US Dollars, recorded for this product. New Zealand, Ireland, Netherland, Belgium, and Germany are the leading nations in the export of butter products. India's export of butter contributes only 0.8 per cent of the total export, ranking 16th. The major destination for the Indian butter product is Egypt, UAE, Turkey, Morocco, and Bahrain.

Table 3.9: Butter Export Time series data

Butter (04051000)		
	Quantity (MT)	**Value (Rs. Crore)**
2007 - 08	4491.82	57.6
2008 - 09	8174.51	116.47
2009 - 10	1275.52	19.99
2010 - 11	5496.69	96.49

Butter (04051000)		
	Quantity (MT)	**Value (Rs. Crore)**
2011 - 12	1742.83	35.72
2012 - 13	988.27	24.88
2013 - 14	3210.48	74.88
2014 - 15	2890.56	80.89
2015 -16	1713.92	54.02
2016 - 17	1401.39	50.17
2017 - 18	7963.83	244.11
2018 - 19	31028.21	807.64
CAGR %	**6.70**	**16.14**

Source: APEDA, 2020

3. Ghee/Melted Butter (04059020): Global export transaction worth of 2,747,247 thousand US Dollars, reported in the year 2018. New Zealand, Netherland, Belgium, Germany, and the United Kingdom were the leading exporting market for this product. Indian export of Ghee/Melted Butter product represents 3 per cent of the global export, ranking 7th. Major importing countries for the ghee/melted butter exported by India were the United Arab Emirates, the USA, Saudi Arabia, Australia, and Oman.

Table 3.10: Ghee/Melted Butter Export Time-series data.

Ghee/Melted Butter (04059020)		
	Quantity (MT)	**Rs. Crore**
2007 - 08	3701.55	55.27
2008 - 09	4213.82	68.54
2009 - 10	3084.23	72.19
2010 - 11	4390.82	91.25
2011 - 12	5346.3	142.05
2012 - 13	5052.63	142.26
2013 - 14	5172.58	163.97
2014 - 15	5707.64	210.61
2015 -16	5594.4	220.96
2016 - 17	6137.02	259.43
2017 - 18	7844.51	354
2018 - 19	10468.43	458.75
CAGR %	**8.48**	**20.26**

Source: APEDA, 2020

4. Butter Oil (04059010): New Zealand, Netherland, Belgium, Germany, and the United Kingdom were the leading exporting market for the trade of butter oil products in the global market. India recorded a Global transaction of 13,598 tons ranking 7th among other leading countries for this product. It is evident from the data available that, the export of this product has gained attention just recently since 2018. Major importing countries for Indian Butter oil are Thailand, Egypt, Bangladesh, Nigeria, and the USA.

Table 3.11: Butter Oil Export Time-series Data

Butter Oil (04059010)		
	Quantity (MT)	**Value (Rs. Crore)**
2007 - 08	872.86	12.97
2008 - 09	4644.01	77.97
2009 - 10	214.18	3.59
2010 - 11	1861.66	38.87
2011 - 12	561.27	13.13
2012 - 13	0.37	0.01
2013 - 14	211.41	7.14
2014 - 15	0.53	0.03
2015 -16	17.4	0.43
2016 - 17	18.42	0.71
2017 - 18	176	6.81
2018 - 19	815.2	28.47
CAGR %	**-25.79**	**-18.59**

Source: APEDA, 2020

4

Value Addition in Milk

In monetary terms, value-added' per unit of the product is the margin between its selling price and cost of production. However, in the dairy and food industry, value addition refers to, extra, features of a product that are beyond the expectations of a consumer at little or no extra cost. In this context, the addition of value to milk can involve monetary value added to a liter of milk by turning it into milk products, enhancing the convenience of quality, shelf life, or health characteristics using processing and packaging interventions.

4.1. Conversion of Milk into Conventional Products

It is an age-old tradition to add value to milk by processing it into milk products. The primary aim was to increase the shelf life of milk that would otherwise be highly perishable. With inadequate facilities to preserve liquid milk fresh in the rural areas' conversion of milk into a variety of milk products would be an ideal option to effectively preserve the huge quantities of milk procured during the flush season' In the processing of milk products such as cream, butter, cheese and ghee, by-products such as skim milk, buttermilk, whey, and ghee residue are obtained' Many of the dairy plants, particularly in India' are usually confronted with the problem of by-products utilization, especially that of whey and ghee residue in an economical manner. Membrane separation processes have presented new possibilities for the production of newer intermediate products from dairy by-products that can be used for value addition in dairy and food industries.

The traditional dairy products were mainly manufactured by the unorganized sector where the major manpower resource was from the family and the raw material was their input. It should be noted that the unorganized sector utilizes a low level of technology and also has a low benchmark in quality control measures, but the value addition from this sector is of great economic significance.

4.2. Development of convenience Dairy Products

With the rising income changing lifestyles, growing women workforce, and the growing number of singles, there is a sharp demand for minimally processed, and processed and packaged foods. Convenience, ready to eat, ready-to-reconstitute, and other foods are gaining popularity as a result. A range of convenience dairy products such as ready-toreconstitute blends, instant mixes, to provide convenience to domestic consumers and the Indian diaspora abroad. Retort processed reduce, ready-to-drink payasam, etc., have been produced. Although most of these convenience dairy products are commercially available and highly popular among consumers, a large number of region-specific convenience dairy products are yet to be developed.

4.3. Development of Milk Products with Enhanced Health Attributes

Traditionally, food products have been developed for taste, appearance, value, and convenience of consumers. The development of new products to confer health benefits (functional foods) is a relatively new trend and recognizes the growing acceptance of the role of diet in disease prevention and treatment. Some of the methods to enhance the health benefits of conventional dairy products are broadly classified as micronutrient fortification; probiotics, prebiotics, and symbiotic dairy foods; dietary fiber fortified dairy products; dietetic dairy products; phytochemical enriched dairy products; omega-3 and other fatty acid fortified dairy products.

4.4. Novel Concepts in Product Development

Emerging concepts have changed the way new products are conceptualized and developed. Milk is considered a unique source of nutrients, but it lacks some micronutrients (iron, copper, and certain vitamins) and dietary fiber. There is therefore a strong need for appropriate sources, to supplement milk with essential micronutrients as well as healthpromoting components. However, the challenges for making composite dairy products include: overcoming the issue affecting the quality attributes, and processing parameters.

4.5. Prebiotic & Probiotic Dairy Product:

There are different nutritional and medicinal properties of fermented foods and beverages. In assessing the positive health effects of fermented milk and related products, lactic acid (LAB) bacteria play a major role. Use of the L. Bifidobacteria spp. and acidophilus in probiotic milk foods are known. At the time of consumption, cultured products sold with any claim for health benefits should meet the requirement of a recommended minimum number of more than 10^6 cfu/g. Yogurt is being redefined as a food carrier for probiotics. Several food powders are made, such as yogurt powder and curd (Dahi) powder, taking into account the number of organisms that live after drying in the product. Thanks to their taste and fragrance and high nutritional value, these foods, drinks, and powders are highly desirable to consumers. (Shilby *et al*, 2013).

4.6. Emerging Technological Interventions

Micro- and nano-encapsulation, Supercritical Fluid (SCF) extraction, non-thermal processing techniques, etc. are rapidly gaining importance for developing value-added dairy products. Micro- and nano-encapsulation techniques have been successfully employed in the past for the controlled delivery of sensitive bioactive compounds and food ingredients in processed dairy foods. Super Critical Fluid extraction can potentially be used for the extraction of phospholipids from milk fat globule membrane, fractionation of triglycerides in ghee, and removal of triglycerides and cholesterol from cheese and fatrich dairy products. Non-thermal processes are food preservation methods to inactivate spoilage and pathogenic microorganisms at temperatures below those used for thermal pasteurization without significant changes to flavour, colour, taste, nutrients, and functionalities. They involve the use of high electric power, high pressure, high intensity of light/radiation, and microfilters.

4.7. Advanced Packaging Systems

The primary functions of a package are to contain, protect, communicate, and provide convenience. The novel food packaging systems have changed the way dairy and food products are packaged for attaining better aforesaid functions. Some of the techniques that gained wide acceptability about food and environment protection include Ultra High Temperature (UHT) processing, modified atmosphere, and active packaging, biodegradable films and edible coatings, bio-nano composite films, etc. UHT processing is extended to preserve milk-based beverages such as lassi, flavoured milk, etc. Presently, dairy products such as a variety of cheese, some indigenous dairy products such as paneer, peda, and brown peda are preserved using the modified atmosphere packaging technique. Growing interest to protect our environment has led to the development of a large number of eco-friendly and biodegradable packaging materials. However, their commercial food and dairy applications are awaited.

In the continued endeavour to facilitate diversification of product portfolio and value addition for the dairy cooperatives, a process has been developed to manufacture Shrikhand-like products without whey drainage. This process not only addresses the environmental issues arising from whey discharge but also improves the nutritive value by retaining the goodness of whey proteins, minerals, and other valuable water-soluble nutrients.

Freeze drying is crucial to successful Ready to Use Culture (RUC) manufacture. It is a time-dependent activity, constituting almost half of the time of production that influences the survival of the RUCs. Hence, the modification was made in the process to address these to achieve a reduction in freeze-drying time without compromising on the survival and activity of the organisms. A laboratory technique has been successfully developed for the production of yogurt RUC. Of the one hundred isolates of lactic acid bacteria, from indigenous fermented products collected from various corners of the country, four promising isolates have been picked up for RUC development. Non-RUC propagative type starter cultures for the production of dahi, mishtidahi, and lassi were supplied to Dimapur Milk Union, Nagaland, and Balaji Dairy, Tirupati Andhra Pradesh, during the year.

The value-added products overall contribute to ~35-40 per cent of the total dairy market in India and commodity products together contribute to almost ~65 per cent of market share. Furthermore, within the value-added segment, largest product category is ghee, having a market share of about 15-18 per cent in the overall dairy market. While loose packets of curd are available locally, a key characteristic of emerging value-added products like UHT milk, flavored milk, low-fat curd/yogurt, cheese and whey is that 100 per cent of these products are sold through organized market. The value-added products market is under-penetrated, thus having tremendous scope for the growth and is expected to grow at much faster rate as compared with the commodity market (Sabnavis *et al*, 2020).

4.8. Demand Drivers

Dual income households, rapid digitization of commerce (e-platforms), increasing disposable incomes growing urbanization, changing consumer lifestyles, and other demographic shifts are driving demand for processed or value-added dairy products. Additionally, the expanding organized retail sector is driving value-added dairy product sales with 15-20 per cent annual growth. For the second and third-tier markets, smaller packaged dairy products are being actively promoted to deepen the penetration of a brand and increase the volume in the country. More based on the untapped opportunities for growth in semi-urban and rural regions is the growing number of players in the dairy sector.

Although our country is the largest milk producer in the world, the conversion of milk into various value-added products is very limited, especially in the organized sector. With the increasing advancements in dairy foods processing and food ingredient science, great scope exists for formulating value-added dairy and food products without compromising with the sensory and overall quality. However, maintaining the ideal quality of perishable agri-food such as milk and milk products are governed by several factors in the supply chain. Understanding the critical issues in the dairy supply chain, including prospects for value addition and measures to address the same, would make the dairy business a sustainable venture. The partnering of the public sector with the nongovernmental agencies may serve as a key move towards the comprehensive development of the nation and its people as a whole. The logical need for partnership in cooperatives and state-owned dairy industries with the private sector is mainly due to the inability of the public sector or government agencies to provide quality goods and services to the public on its own effectively and justifiably.

It has to recognize that Indian consumers, in general, have a clear preference for liquid milk in various forms of indigenous products, and value addition is done to produce diversified milk products (skimmed milk, double toned milk, toned milk, standardized milk, high-fat milk, flavoured milk, etc.). Similarly, the manufacturing of various indigenous products, ethnic products, and fresh milk products adds value to the base product, i.e., milk. It is important that the manufacturing of these products is standardized and hygiene, as well as quality norms, are adhered to for the safety of human health and quality of life.

Buoyancy is a favourable factor in market demand for milk. For a prolonged period, this buoyancy was felt to suggest a high potential for development. Such innovations will activate additional criteria for value-added goods. Therefore, investment in improving efficiency, milk procurement, milk chilling, processing, and marketing through the cool chain will be needed by the industry.

5

Impact of COVID 19 on Dairy Sector

Livestock is considered one of the essential variables in the revenue of a farmer. The sector (excluding fisheries) contributes 30 per cent to India's agriculture gross domestic product, and it is valued at more than Rs. 9 lakh crores. India is also one of the largest producers of milk in the world; around 70 million rural households are engaged in professions linked to dairy. India is also the second-largest exporter of bovine meat after Brazil. This sector provides a substantial income to farmers and supports the economy by generating enough demand in the rural and semi-urban regions. This animal husbandry sector comprising dairy and poultry is extremely valuable for farmer's income security, and Covid-19 has the potential to disrupt this sector.

In developing countries such as India, where dairy farming is one of the most important sources of livelihood for 60 million rural households, 80 per cent of whom are landless, small and marginal farmers who depend on dairy for daily income, the crisis is extreme. Thus, even a small decline in the growth of the dairy sector significantly affects a large chunk of our population which are already vulnerable due to low income, poor savings, and lack of livelihood alternatives. Small production in multiple households leads to more reliance on efficient transportation for procurement. In addition to this, the Indian dairy sector is still dominated by the unorganized sector which handles more than 50 per cent of the marketable surplus. Milkman and vendors collect a small quantity of loose milk from different households and then supply it to urban consumers, roadside eateries, hotels, and sweetshops. Closure of shops and restrictions on transportation has ruthlessly disturbed this arrangement. Even in the formal sector out of the total milk processed 65 to 70 per cent is sold as liquid milk which is immediately vended after packing and only 30 per cent is processed into other dairy products like ghee, butter, cheese, curd, ice-cream, milk powder, etc. Thus, the formal sector also lacks sufficient infrastructure and storage facilities for processing and storing surplus milk after value addition.

The Pan-India lockdown has impacted liquid milk consumption by around 30-35 per cent with the closure of the HoReCa (Hotel, Restaurants, and Catering) sector, QSR (Quick Service Restaurants, Casual Dining), unorganized sector, as well as sweetmeat shops. Further, apart from the sweetmeat and unorganized segment, the others are expected to require a minimum of 6-9 months to bounce back to pre-Covid level demand, post lockdown. (FICCI – 2020).

The loss of demand has led to pressure on milk procurement prices. For the Industry, given the COVID-19 situation, there is an oversupply of milk for (reduced) demand. Thus, industries procuring milk to covert to SMPs are running at 80-100 per cent capacities as of date against 30 to 50 per cent in normal lean milk production months (April-September, 2020). This excess production would contribute to the domestic collapse of SMP prices. Also, the Ice Cream Industry's SMP intake would be small, as the industry is projected to be dramatically affected. Meanwhile, the global prices of SMP have also crashed from USD 3000/MT in Jan 2020 to USD 2514/MT in April 2020 which is expected to fall further— thus impacting export opportunity of excess SMP.

5.1. Impact of Covid-19 Dairy Supply Chain

Supply chains are highly integrated and a minor effect on any of the processes significantly affect all the other processes and actors in the chain. This is why the effect of movement restrictions during the lockdown is being felt along the entire supply chain. The following paragraph discusses some major effects of lockdown and supply chain disruptions on dairy production and processing based on interaction with some dairy farmers and small dairy processors.

a) Dairy Production- Dairy farmers are facing problems in procuring inputs as well as in disposing of their output. Ensured supply of inputs is a sine-qua-non for any efficient production process and this is one of the components which have suffered heavily due to lockdown restrictions. Feed and fodder accounting for approximately 60 per cent of the total cost of milk production are the major inputs in dairy farming followed by labour and veterinary services. A balanced ration for cattle includes an optimum proportion of green fodder, dry fodder, and concentrate. With increased commercialization and decreasing landholding size, dairy farmers are now more dependent on the market for meeting their requirements of feed and fodder which are sourced from different parts of the country. This is why the imposition of lockdown restrictions significantly affected their availability. The farmers were facing more problems in getting cattle feed than fodder. This is mainly because the cattle feed industry has temporarily shut down its production units due to the irregular supply of raw materials, labour shortage, and transportation issues. Recently, the National Dairy Development Board (NDDB) started reformulating the cattle feed by substituting the raw materials in short supply with locally available ingredients.

Labour is another major input posing a severe problem, particularly for commercial dairy farms as most of the migrant laborers have either returned to their home towns or are not able to reach their workplace. Though there was not much problem regarding the availability of medicines for animals but veterinary services like artificial insemination has been suffering badly. The farmers will have to face the aftermaths of missed artificial insemination for a longer time. The other components bearing the big brunt are the flow of inputs and credit.

If we look into the cost aspect, there was more or less a rise in the prices of all the inputs. The biggest hike was observed in the prices of cattle feed which increased by Rs. 150-200 per quintal whereas the dry fodder prices increased by 80-100 rupees per quintal. Prices of feed and fodder were on the upswing during the last year also due to adverse climate and floods in various parts of the country, Covid-19 just made it worse for the dairy farmers. Coming to the output, farmers reported that daily milk production has declined by approximately 10-15 per cent due to the limited availability of feed and fodder. The average daily sale of milk has also declined. Commercial dairy farms reported an 8-10 per cent dip in their daily milk sales. Dairy farmers who were directly delivering the milk to consumer door-steps in the near-by houses have been affected less as compared to the ones supplying it to the dine-in outlets or private processors. Due to a fall in demand, farmers received lower milk prices. Estimation based on newspaper reports indicated that the average milk prices have declined by 35-40 per cent for most of the dairy farmers across the country. During our interaction, we found that some of the commercial dairy farmers were having a strong and close customer network, and for them, though there was no change in the prices received for milk but they faced severe issues in milk delivery which affected the total sales. Transportation particularly inter-state movement posed a great problem and much of the milk was spoiled during this.

b) Dairy Processing- The dairy processing industry in India is quite small in comparison to the huge quantum of milk produced in the country and the total share of cooperatives and private processing units is hardly 12-15 per cent. Most of the milk moves in the form of liquid milk and a smaller proportion are converted into other dairy products. Though the major dairy cooperatives like Amul were able to ensure maximum procurement most of the private players faced problems in procurement and proper transportation of milk. Another major issue was a shortage of packaging material and limited processing capacity. The problem was relatively less in the availability of other inputs like labour, sugar, fuel, and lubricants.

The key concern for dairy processors was a fall in demand due to the shutdown of restaurants, hotels, sweet shops, roadside eateries, offices, and schools owing to which they had to keep their plant idle. Demand for liquid milk fell sharply than other products like ghee, paneer, curd, etc. Some of the dairy processing units reported an increase in the demand for paneer during the lockdown period. There was not much increase in the input cost for processing units. Only transportation cost showed an upward trend for them while some of them got milk at a lower cost. Thus, the fall in demand remained a major culprit for them. Along with this, they were not able to get timely payments from their customers which also reduced their efficiency.

Covid-19 has opened new avenues in terms of packaged branded dairy products for the household consumer groups. This is prompting dairies to innovate with products suitable for home consumption.'Haldi ice cream' was launched by Amul, which has simple ingredients such as sugar, pepper, and dry fruits such as dates, almonds, and cashews. Haldi's ice cream consists of 0.8 per cent turmeric, 0.25 per cent black pepper, 5.5 per cent dates, 2 per cent almond and cashews, and 6.2 per cent "honey ripple syrup." Haldi or Turmeric has antibacterial and anti-inflammatory properties. The Gujarat Cooperative Milk Marketing Federation (GCMMF), which markets all milk and dairy products under the Amul name, has launched 'Haldi ice cream' to improve customer immunity, with all the benefits it claims for turmeric.

The Kerala Cooperative Milk Marketing Federation (KCMMF) and Milma have launched two new products, Milma golden milk and Milma golden milk mix, which is spiced with ginger, turmeric, and cinnamon. It is said that the latest editions have elevated nutritional value, improve immunity, and help combat pathogens. The products were launched with the support of the Indian Institute of Spices Research, Kozhikode, which established that by properly processing and segregating the properties and later mixing them with proteins in pure milk, bioactive spices would help to improve the immunity system (Dairy Times, 2020).

Fluid Milk Prices

Dairy cooperatives set farm gate fluid milk prices and are generally benchmarks for private actors who procure milk in the region. The dairy cooperatives consider factors such as feed and fodder cost increases, in addition to other inputs when revising farm gate milk prices. For example, cow milk procurement price which was upwards of Indian Rupee (INR) 30-31/liter in March 2020 ($0.40 to $0.42) was reported to have slid to INR 18-20 per liter ($0.24 to 0.26/ liter), but by mid-July had recovered to INR 24-25/ liter ($0.32 to $0.33 per liter) in August due to some demand amid government-eased COVID-19 restrictions. To offset some of the loss in profits, some state governments provide subsidies via cash transfers to improve producer's margins.

Being an essential commodity with its supply deemed an essential service, milk availability has been steady through the lean period. The flush season (October 2019 through March 2020) was extended through April- May 2020 which partly ensured milk availability when the nationwide COVID-19 lockdown started on March 24. Starting April 2020, the milk wholesale price index (WPI) rose just 0.5 per cent to INR 152.1 in August 2020, while that of ghee, butter, and powdered milk fell three, ten, and 11 per cent, respectively. India's Fodder Price Index, which initially stood firm, also softened five per cent with improved availability owing to good monsoon rains. However, the Cottonseed and Mustard Oil Cake Indices grew seven per cent due to tight supply and strong animal feed demand (USDA GAIN, 2020).

The Indian national average consumer retail price of milk as reported from the GOI's Department of Consumer Affairs ranged from INR 45.21 (USD 0.61)/ liter in January 2020 to INR 47.0/liter through mid-September. In northeast India, milk was retailed at a premium of INR 10/liter and in the eastern region was sold at a discount of INR 5/liter as compared with the Indian average.

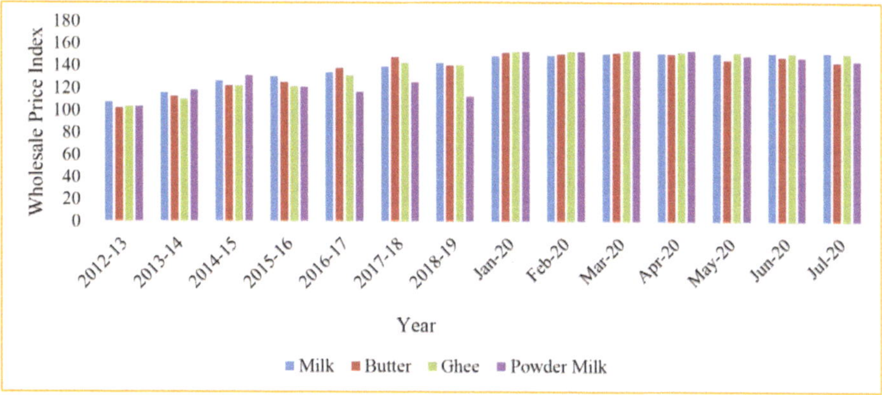

Fig. 5.1: India's Wholesale Price Index for Milk, Butter, Ghee, & Powder Milk
Source: Office of The Economic Adviser

There is growing health consciousness and subsequent demand for healthier food products, especially in 2020 amid the COVID-19 global pandemic. Consequently, consumers, including those with higher disposable income will continue to demand valueadded dairy products including fermented yogurts, turmeric, and nutrient-based health drinks that feature milk, buttermilk, processed cheese (e.g., mozzarella, spread, flavored and spiced), and ice cream. Besides, the demand for ghee and butter will continue to be robust especially as ghee is one of the most consumed value-added dairy products in India.

During the national lockdown, surplus milk purchased by dairy cooperatives was converted to clarified butter, fat, and milk powder and remained unsold amid poor purchases from bulk and institutional buyers. Household demand however picked up fast, due to the rise in a new consciousness toward health, hygiene, and nutritional food products. This behavioural change kept sales moving, while only partly compensating for poor sales in the hotel, restaurant, and institutional segment. Following the national lockdown and gradual reopening of the economy, the demand for dairy and milk products is expected to resume. With the flush season approaching, milk supply will increase and should match concurrent seasonal and festive demand. Without this demand, higher supplies will negatively impact the milk procurement price, and the differences with consumer retail pricing will only grow wider.

5.2. Impact of COVID 19 on Exports and Imports

India exports mostly value-added dairy products such as butter, butterfat, infant food preparations, cheese (Cheddar and Colby), milk powder, lactose, dairy spreads, and milk albumin. Non-fat Dried Milk (NFDM) exports in 2021 are expected to rise to 20,000 MT (close to the five-year average) from an estimated 6,000 MT in CY 2020, with increasing domestic milk production in 2021 and a likely resumption of demand from overseas buyers. This volume could further rise by 25 per cent if global trade resumes to near-normal levels before the global pandemic. January-June 2020 trade data indicates that India exported skimmed milk powder (SMP) worth $1.9 million on shipments totalling 500 MT, which is 89 per cent below the corresponding period in 2019 due to lower-thanexpected purchases (pandemic-induced supply disruption) from regular buyers such as Bangladesh (the largest buyer), Afghanistan, the United Arab Emirates (UAE), Turkey, and Bhutan. In terms of ranking (January-June 2020), NFDM, SMP were a combined fifth in largest exports (value basis) after butter, butterfat, infant food preparation, and cheese. The monthly average NFDM sales declined from its peak of 300,000 MT in January 2020, to spike intermittently in May/June and dip to its lowest (less than 20,000 MT) average monthly sales in July. Industry sources indicated some trade resuming in August 2020 (USDA GAIN, 2020).

Concurrently, total butter exports will rise from 16,000 MT (revised to reflect current trade estimate and 64 per cent below last year) in 2020 to 25,000 MT for 2021, due to modest demand from traditional and new buyers. Trade data from January-June 2020 indicate total butter sales at 2,200 MT worth $10.2 million, almost 90 per cent below the corresponding period last year. Amid the global pandemic, principal buyers such as Turkey, Egypt, UAE, Morocco,

Bahrain, Saudi Arabia have also reduced their purchases. Incidentally, the United States and Bhutan increased their purchases from India by 41 per cent and 40 per cent respectively, with U.S. imports volumes at 910,000 MT and Bhutan at 791,275 MT (USDA GAIN, 2020).

The Ministry of Finance presented India's annual budget for IFY 2020/21 on February 1, 2020 (April 1, 2020, to March 31, 2021). With each annual budget, the GOI has an opportunity to review the structure of import tariffs and, through its annual budget, propose and implement changes to India's Customs and Central Excise Board regulations and duty rates. Permanent adjustments in the tariff system are rarely considered for food and agricultural products and are regularly reviewed; however, the recent announcement was solely for BCD rises for several agricultural goods.

Table 5.2.1: New Custom Tariffs by Govt of India on Dairy Products

HS Code	Product Description	Old BCD	New BCD	Total Imports in CY 2019- (Jan – Nov)
04041010 & 04049090	Whey under referred HS Codes	30 %	40 %	Negligible
0405	Butter, Ghee, and Butter Oil	30 %	40 %	$ 2.4 million
04069000	Other Cheese	30 %	40 %	$ 4.3 million

Source: USDA GAIN, 2020

In addition to raising the basic duty rate, the notification also omits a preferential duty rate for imports of products under the tariff rate quota. However, the Ministry of Finance sources report that Tariff Rate Quote (TRQ) tariff rates on these products have been omitted from the tariff table shown below.

Table 5.2.2: Changes in Dairy Products Tariff Rate Quotas by GOI

HS Codes	Product Description	Old Tariffs	Preferential tariffs omitted	Total imports in CY 2018
04022100	Milk Cream not containing added sugar or sweetening matter	15% up to an aggregate of 10,000 MTs under TRQ; 60% out of TRQ	15% in TRQ omitted; 60% tariff for all imports	$0.1 million

Source: USDA GAIN, 2020

On 24 July 2020, the FSSAI extended the Stakeholder Implementation Date to implement Bureau of Indian Guidelines (BIS) standards relating to commercial feed/feed materials intended for meat and milk-producing animals. The new compliance date is January 1, 2021, and as they deal with the current COVID-19 pandemic and subsequent lockdown orders issued by the GOI, the timeline has been extended to accommodate requests from sector stakeholders.

On October 14, 2020, the Department of Agriculture, Farmer Welfare and Cooperation, the state government of Gujarat introduced a resolution to approve export subsidies for skim milk powder (SMP), scheduled to begin from November 1, 2020, to April 30, 2021. The resolution indicates the Gujarat government will provide total assistance of Rs 150 crores (US$20 million) for six months to Gujarat Cooperative Milk Marketing Federation Ltd. The export subsidy will entail Rs 50 per kilogram of SMP exported by the milk cooperatives if the FOB price is Rs 180. The total amount of SMP that can be exported under the scheme is 30,000 metric tons (USDA GAIN, 2020).

6

Dairy Schemes & Policies

The GOI is pushing to strengthen infrastructure for the production of quality milk, as well as for procurement, processing, and marketing of milk and milk products through the following Dairy Development Schemes:

6.1. Dairy Entrepreneurship Development Scheme (DEDS)

Dairy Entrepreneurship Development Scheme (DEDS) was started in September 2010. This scheme is being implemented through NABARD which provides financial assistance to commercially bankable projects with loans from Commercial, Cooperative, Urban and Rural banks.

Objective of the scheme

To generate self-employment and provide infrastructure for the dairy sector, to set up modern dairy farms and infrastructure for the production of clean milk, to encourage heifer calf rearing for conservation and development of good breeding stock, to bring structural changes in the unorganized sector so that initial processing of milk can be taken up at the village level, to upgrade traditional technology to handle milk on a commercial scale, to provide value addition to milk through processing and production of milk products.

Since inception, under DEDS, NABARD as implementing agency has disbursed Rs. 1712.37 Crore as back-ended capital subsidy (as of 31.12.2019) to 387872 beneficiaries out of which are 1,02,078 women beneficiaries and 65,782 SC/ST beneficiaries.

Components of the Scheme

- Establishment of the small dairy unit from 2 to 10 milch animals, Rearing of heifers (up to 20 calves),
- Vermi-Compost,
- Purchase of Milking Machines,
- Milko testers & BMCs (up to 5000 liters capacity),

- Purchase of Milk Processing equipment for the manufacture of indigenous milk products,
- Transportation & Cold Storage facilities,
- Establishment of private veterinary clinics,
- Setting up of Milk Parlour for enhancement of milk production, Procurement, Cold chain, and Transportation facilities,
- Processing and Marketing of milk & milk products.

Since inception, under DEDS, NABARD as implementing agency has disbursed Rs. 1712.37 Crore as back-ended capital subsidy (as of 31.12.2019) to 387872 beneficiaries out of which are 1,02,078 women beneficiaries and 65,782 SC/ST beneficiaries.

Achievements of DEDS

Since inception, under DEDS, NABARD as implementing agency has disbursed Rs. 1712.37 Crore as back ended capital subsidy (as on 31.12.2019) to 387872 beneficiaries out of which are 1,02,078 women beneficiaries and 65,782 SC/ST beneficiaries.

6.2. National Dairy Plan I (NDP I)

National Dairy Plan Phase-I (NDP-I) is a scientifically planned multi-state initiative to increase the productivity of milch animals and thereby increase milk production to meet the rapidly growing demand for milk through scientific breeding and feeding and to provide rural milk producers with greater access to the organized milk processing sector.

NDP-I is an externally aided project with a total outlay of Rs 2242 crore comprising Rs 1584 crore as International Development Association assistance, Rs 176 crore as Government of India share, Rs 282 crore as share of EIAs for the period 2011-12 to 201819.

Objectives

- Increase the productivity of milch animals and thereby increase milk production to meet the rapidly growing demand for milk.
- Provide rural milk producers with greater access to the organized milk-processing sector.

Coverage

NDP-I is being implemented in 18 major milk-producing States, viz. Andhra Pradesh, Bihar, Chhattisgarh, Gujarat, Haryana, Jharkhand, Karnataka, Kerala, Madhya Pradesh, Maharashtra, Odisha, Punjab, Rajasthan, Tamil Nadu, Telangana, Uttar Pradesh, Uttarakhand, and West Bengal. These States account for more than 90 per cent of the country's milk production. However, the benefits from the project are accruing across the country.

Funding Pattern

The pattern of funding under the scheme is 100 per cent grant-in-aid for nutrition and breeding activities and in the case of village milk procurement systems, 50 per cent of the cost of capital items is being contributed by the End Implementing Agencies.

Major Components Funded Under the Scheme

Breed Improvement: Production of High Genetic Merit (HGM) cat t le and buffalo bulls, Strengthening of "A" and "B" graded Semen Stations, Pilot Model for Viable Doorstep AI delivery Services.

Animal Nutrition: Ration Balancing Programme, Ration Balancing Programme, Fodder Development Programme.

Village Based Milk Procurement System: Strengthening and Expanding Milk Procurement System at Village level.**2**

Achievements

Under NDP I, 577 sub projects with total grant assistance of Rs. 1759.97 Crore were approved and the utilisation till March 2020 was Rs. 1638.14 Crore. The total utilisation under NDP I including Rs. 405.72 Crore as EIA contribution and Rs. 194.47 Crore as NDDB and its subsidiaries contribution was Rs. 2238.33 Crore.

Under NDP I, 2456 High Genetic Merit Bulls (98 % of the target) were made available to A & B graded semen stations across the country which propelled the production of quality disease free semen doses to about 119 Million doses during the year 2018-19 (119 % of the target). The project also contributed towards lowering the cost of feeding per kg of milk by about 11.62 per cent as against project target of 7 per cent which resulted in increase of net daily income of milk producers by about Rs. 8000 per lactation of 305 days. Market access was provided to more than 16.88 Lakh milk producers. Overall, the project covered around 59 Lakh beneficiaries across 97,000 villages. Greater emphasis on transparency and fairness of milk procurement at the village level led to installation of more than 29,500 AMCUs/ DPMCUs and 4171 BMCUs under the project. Further, the project was instrumental in spearheading women empowerment in the dairy sector with more than 7.65 Lakh women (45 %) members enrolled as members at village level dairy cooperative societies.

Under NDP I, an innovation fund was established to promote new/ innovative activities related to Dairy Sector. Details of the activities approved under New/ Innovative is provided below:

A. Genetic Improvement and Animal Health Management

1. Developing genomic selection methodology for various cattle breeds in India: Use of "INDUSCHIP" for genotyping recorded animals of Indigenous cattle breeds & their crosses and developed appropriate model for estimation of Genomic Breeding values of HF crossbred, Jersey crossbred and Gir bulls.
2. Developing and validating genotyping microarray chip for buffaloes for genomic selection: First of its kind medium density microarray chip "BUFFCHIP" was developed and close to 4500 Murrah buffaloes were genotyped using BUFFCHIP.
3. IBR control through inactivated marker vaccine: As per our evaluation study, more than 98 per cent of the IBR vaccinated animals housed in endemic environment remain uninfected for IBR during the study period
4. Promotion of renewable energy resources and ICT tools: Flexi Bio Plants: Total 1500 flexi-biogas plants of 2-3 cubic metre capacity were provided on 50 per cent grant basis as a pilot project to promote affordable & easy to use clean energy alternative at village level to 32 Milk Unions across the country. The interventions are expected to contribute an annual reduction of about 817 Metric Tonnes of GHG emissions (in terms of CO2 equivalent) owing to the replacement of LPG.

Solar Rooftop: Total 61 solar rooftop solar panel of 2KW capacity were installed at DCS's on a pilot basis to reduce dependency on grid connected power supply.

Data management and use of ICT: Data logger systems for providing real time data on various parameters related to BMC functioning and a total of 69 Data loggers were installed on pilot basis covering 69 Milk Unions/ PCs across the country.

B. Feed and Fodder

Pilot to test the effectiveness of the use of Moringa as a fodder crop was undertaken and the data is being collected for analysis and study.

The implementation of NDP I ended on 29 Nov 2019 as per the project closure date of World Bank and the financial closure was done in March 2020. By the close of the project, NDP I had achieved the project development objectives and the Intermediate Result Indicators. The activity wise physical progress is provided as Table 6.2.

The World Bank has published the Impletion Completion and Results Report (ICRR) of NDP I and the outcome rating for NDP I is "Highly Satisfactory" which is the highest rating in the World Bank funded projects.

Table 6.1: Activity Wise Financial Progress – NDP – I

Activity	No. of Approved Sub Projects	Amount in Rs. Crore	
		Grant Assistance	Fund Utilisation till Mar 2020
Animal Breeding	62	664.53	618.32
Progeny Testing Programme	14	203.54	197.45
Pedigree Selection Programme	9	34.19	32.71
Strengthening of Semen Stations	28	294.98	287.01
Pilot AI Delivery Services	4	77.28	75.97
Import of Bulls	1	27.93	12.93
Import of Embryos/ Semen/ BPTIE	6	26.61	12.23
Animal Nutrition	169	306.21	297.46
Ration Balancing Programme	117	234.24	228.01
Fodder Development	52	71.97	69.45
Village Based Milk Procurement System	243	686.23	640.48
Sub Total	474	1,656.97	1,556.26
Project Management & Learning	103	103.00	81.89
Total	577	1,759.97	1,638.14

Source: NDDB, 2020

Table 6.2: Physical Progress – NDP I

Indicators	Unit of Measure	Baseline	Cumulative	
		2012	EoP Target as per PIP	EoP Actual
PDO level indicators				
Milk production/animal	Liters/day	5.03	5.53	5.8
Proportion of "in-milk" female animals to adult female animals	%	63	66	67

Indicators	Unit of Measure	Baseline	Cumulative	
		2012	EoP Target as per PIP	EoP Actual
Proportion of total milk sold to total production	%	65	65	66
Share of milk sold to the organized sector (as a share of production)	%	45	56	59
Intermediate Result Indicators				
Number of genetically improved bulls				
produced through progeny testing program*	Numbers	25	2025	2185
Number of genetically improved bulls produced through pedigree selection*	Numbers	0	500	271
Number of semen doses produced	All India Million Doses (Annual-2018-19)	56	100	119.07
Number of villages covered by AI	Numbers	0	24000	12322
Number of AIs done annually during 2018-19	Million	0	3.82	0.78
AI Conception Rate	%	30	45	44
Villages covered und. Compoer RBP	Numbers	400	40000	33374
Animals covered by RBP	Lakh	0.1	27	28.65
Per cent reduction in cost of feeding per Kg of milk for animals covered by RBP	%	0	7	Coop:11.8 PC: 8.29
Number of trained mobile AI technicians (MAITS) in the project	Number	0	3000	1330
Number of trained Local Resource Persons (LRPs) in the project	Number	400	40000	31148
% reduction in methane release per				
kilogram of milk produced by animals covered under RBP	%	0	10	13.3
Number of additional milk producers organized or enrolled into dairy coop societies (DCS)	Total (Lakh)	0	12	16.88
	of which female (Lakh)	0	3	7.64
	of which Small holders (Lakh)	0	9	11.35
Additional number of villages covered or strengthened for milk collection	Numbers	0	31900	52509

Source: NDDB, 2020

6.3. National Programme for Dairy Development (NPDD)

National Programme for Dairy Development was initiated in March 2014 with a total scheme outlay of Rs.600 crore for period up to the end of 12th Plan i.e., 2016-17. The scheme is aimed at strengthening dairy infrastructure covering village level procurement centres, establishment of bulk milk cooling units, milk processing plants at district & State level and marketing infrastructure for facilitating Milk Unions and Federations to market their products. The scheme with the same objectives and target has been implemented during 2017-18 to 2019- 20.

Objectives

The NPDD focuses on creating/strengthening of infrastructure for Production of quality milk, Procurement, Processing and Marketing of Milk & Milk Products by the State Implementing Agency (SIA) i.e., State Cooperative Dairy Federations/ District Cooperative Milk Producers' Union.

- To Create and Strengthen Infrastructure for Production of Quality Milk Including Cold Chain Infrastructure Linking the Farmer to The Consumer;
- To Create and Strengthen Infrastructure for Procurement, Processing and Marketing Milk;
- To Create Training Infrastructure for Training of Dairy Farmers;
- To Strengthen Dairy Cooperative Societies/ producers Companies at Village Level;
- To Increase Milk Production by Providing Technical Input Services Like Cattle-Feed, And Mineral Mixture Etc;
- To Assist in Rehabilitation of Potentially Viable Milk Federations/ Unions;

Achievements under NPDD

129 projects in 28 States and 2 Union Territory have been approved with the total cost of Rs. 1535.17 Crore (Central Share Rs. 1249.36 Crore) from 2014-15 to 2019-20(till 31.12.2019). A total sum of Rs. 850.63 Crore has been released for implementation of new projects approved under the scheme up to 31.12.2019 (DAHD, 2019).

Physical Progress under NPDD

- 2.65 lakh new farmers were given benefit of membership of dairy co-operative societies.
- 16.89 lakh litres per day new milk processing capacity has been established.
- 731 Bulk Milk Coolers with 8.07 lakh litres chilling capacity installed,8158 Automatic Milk Collection Unit and Data Processing and Milk Collection Unit installed at village level dairy cooperative societies.
- 2667 Electronic Milk Adulteration Testing Equipment and 502 Butyro-Refractometer has been approved for checking adulteration in milk.
- Approval has been granted for strengthening of milk testing laboratories of all 231 Cooperative Milk Unions in the country to check adulteration in milk.
- State Central Laboratory has been approved in 18 States.

DAHD has received 164 proposals from various state governments under NPDD during March 2014-February 2020. As of 25.02.2020, 129 new projects have been approved in 28 States and 2 UTs with a total outlay of Rs 1535 crore (Central Assistance Rs 1249 crore) during 2014-15 to 2019-20. (DAHD, 2019)

6.4. Dairy Processing and Infrastructure Development Fund (DIDF)

The government of India (GoI) launched the Central Sector Scheme – Dairy Processing and Infrastructure Development Fund (DIDF) during 2017-18. The scheme has a total outlay of Rs 10,881 crore comprising Rs. 8,004 crores as a loan from National Bank for Agriculture and Rural Development (NABARD), Rs. 2,001 crores as End Borrower's contribution, Rs. 864 crores as GoI's Interest Subvention and Rs. 12 crores to be contributed by Implementing agencies (NDDB and NCDC) towards project management and learning.

Objectives

The scheme focuses on the creation/modernization/expansion of processing infrastructure and manufacturing facilities for Value Added Products to make operations of milk processing plants more efficient as well as geared on the production of higher Value-Added Products. The project also focuses on setting up chilling infrastructure & installation of electronic milk adulteration testing equipment at the village level.

Implementing Agencies

The project will be implemented by NDDB/ NCDC directly through the End Borrowers such as Milk Unions, State Dairy Federations, Multi-state Milk Cooperatives, Milk Producer Companies, and NDDB subsidiaries meeting the eligibility criteria under the project.

Salient Features

- Under the Scheme, a long-term loan at the rate of 6.5 per cent per annum is provided to End Borrowers for a maximum period of 10 years including a maximum 2 years moratorium on repayment of principal.
- With this investment, the additional milk processing capacity of 126 lakh liter per day, value-added products manufacturing capacity of 59.78 lakh liter per day of milk equivalent, milk drying capacity of 210 MT per day, milk chilling capacity of 140 lakh liter per day shall be created along with the installation of 28,000 electronic milk testing equipment.
- About 95,00,000 farmers in about 50,000 villages would be benefitted under the scheme.

Implementation Status

- Till February 2020, 33 projects have been approved with a total project outlay of Rs. 4059 crores including a loan of Rs. 2722 crore. Further, 4 projects with an outlay of Rs. 399 Crore (Loan Component: Rs. 485 Crore) are in pipeline.
- Till February 2020, NABARD has released an amount of Rs. 1110 crore to NDDB.
- Union Cabinet has increased the interest subvention up to 2.5 per cent.

6.5. Rashtriya Gokul Mission (RGM)

The Rashtriya Gokul Mission has been implemented for the development and conservation of indigenous bovine breeds since December 2014. The scheme is crucial for the upliftment of rural poor as more than 80 per cent of low producing indigenous animals are with small and marginal farmers and landless labours. The scheme is important in enhancing milk production and productivity of bovines to meet the growing demand for milk and making dairying more remunerative to the rural farmers of the country. The scheme is leading to the multiplication of elite animals of indigenous breeds and increased availability of indigenous stock.

Objectives

The scheme is being implemented with the following mandate:

1. To undertake breed improvement program for indigenous cattle breeds so as to improve their genetic makeup and increase the stock.
2. To enhance milk production and productivity of indigenous bovine breeds.
3. To upgrade nondescript cattle using elite indigenous breeds like Gir, Sahiwal, Rathi, Red Sindhi, Tharparkar, Kankrej, and Hariana.
4. Arrange quality Artificial Insemination (A.I) services at farmers' doorstep
5. Bring all breedable females under organized breeding through Artificial Insemination using germplasm of high genetic merit disease free bulls.
6. An increasing population of high genetic merit female population.

6.6. Rashtriya Krishi Vikas Yojana – Remunerative Approaches for Agriculture and Allied Sector Rejuvenation (RKVY-RAFTAAR)

The Rashtriya Krishi Vikas Yojana (RKVY) was launched in 2007 to operationalize the "resolution of the 53rd meeting of the National Development Council (NDC) held on 29th May 2007 that a special Additional Central Assistance (ACA) Scheme is initiated." The twin purpose of the RKVY as conceived was to encourage States to draw up District and State Agricultural Plans and also increase their spending on the sector to reorient agricultural development strategies for rejuvenating Indian agriculture during the Eleventh Plan. The maximum flexibility in selecting interventions and setting state-specific targets was inbuilt in RKVY and hence was preferred by the states.

One objective of RKVY was to incentivize states to increase expenditure on agriculture and allied sectors. State Plan expenditures (excluding RKVY receipts) as a percentage of GDP in agricultural and allied increased from 1.0 per cent in the 10th Plan to 1.4 per cent in the 11th Plan. State Plan expenditures on Agriculture and Allied Sectors (excluding RKVY) have also increased as a percentage of total Plan spending by States, from about 5 per cent during 10th Plan to over 6 per cent during 11th Plan. RKVY was therefore successful in motivating States to pay greater attention to agriculture, besides providing increased Central assistance for the sector. RKVY as assistance was particularly useful for the fund starved Animal Husbandry, Dairying, and Fisheries sectors. Projects amounting to over Rs. 5000 crores were sanctioned under RKVY for these sectors during the 11th Plan, about 20 per cent of the total sanctioned

RKVY projects, and more than spending on DAHD's schemes. This has provided a substantial push to these sectors which accounts for a significant contribution to the Agricultural Gross Domestic Product (GDP).

An Illustrative List of Dairying Related Components Being Funded

i. Semen collection and Artificial Insemination Units/ Production centre

ii. Breeding Farms

iii. Dispensaries/ Hospitals for treatment of animals

iv. Vaccine Production Unit

v. Diagnosis Labs, including Mobile Units

vi. Animal Ambulance

vii. Cold Chain for storing and transportation of frozen semen

viii. Tractor fitted with Fodder Block Machine

ix. Carcass rendering Plant to collect the fallen animals for processing/ utilization in a scientific manner.

x. Modernization of animal slaughterhouses and markets for Livestock/ Livestock products.

xi. Establishment/ Strengthening of cold chain infrastructure for storage of veterinary biological.

xii. Establishment/ Strengthening of Check post/ Quarantine camps for restriction of animal movement, strengthening of animal disease reporting system.

xiii. Vermicompost.

Milk Collection Centres and Infrastructure

xiv. Purchase of milking machines (single/double bucket).

xv. Setting up of milk chilling/bulk milk cooling centres (BMC) along with automatic milk collection units (AMC).

xvi. Setting up/modernization/strengthening of milk processing units Strengthening /expansion of cold storage facility for milk and milk products xvii. Purchase of insulated/refrigerated transport vehicles xviii. Setting up of milk parlour/milk booth.

xix. Strengthening of lab facility in milk chilling/milk processing unit.

xx. Establishment of cattle feed storage go-down.

xxi. Establishment/strengthening of cattle feed plant. xxii. Establishment of cattle shed for milch animals.

xxiii. Setting/strengthening of ETP at milk chilling/ milk processing unit.

6.7. National Livestock Mission (NLM)

For the sustainable and continuous growth of the livestock sector by emulating the success achieved in Dairy and Poultry sectors, across species and regions, the National Livestock Mission was launched in 2014-15 with an approved outlay of Rs. 2,800 crores during XII Plan. This Mission is formulated with the objectives of sustainable development of livestock sector, focusing on improving availability of quality feed and fodder, risk coverage, effective extension, improved flow of credit and organization of livestock farmers/ rearers, etc. with the following four Sub-Missions:

1. Sub-Mission Livestock Development,
2. Sub-Mission on Pig Development in North-eastern Region,
3. Sub-Mission on Fodder and Feed Development,
4. Sub-Mission on Skill Development, Technology Transfer, and Extension.

It broadly covers all the activities required to ensure quantitative and qualitative improvement in livestock production systems and capacity building of all stakeholders. The major outcomes of the Mission envisaged are mainstreaming of livestock rearing as business models and linkages for successful business ventures to achieve 5-6 per cent annual growth rate, optimal utilization of scarce nutritional resources – reducing the gap in demand and availability of fodder, conservation, and improvement of indigenous breeds, higher productivity, and production in a sustainable and environmentally friendly manner, enhanced livelihood opportunities, especially in rainfed areas and for landless, small and marginal farmers, increased awareness, improved risk coverage and better availability of quality animal products to consumers overall socio-economic upliftment of livestock rearers.

6.8. Livestock Health and Disease Control (LH&DC)

To effectively tackle the issue of livestock health, the Department supplements the efforts of the State Governments / Union Territories by way of financial assistance through the 'Livestock Health & Disease Control (LH&DC) Scheme. At present, the scheme has been clubbed Research Institute (IVRI), Izatnagar is functioning as the Central Disease Diagnostic Laboratory (CDDL).

The Disease Investigation Laboratory (Pune), Institute of Animal Health and Veterinary Biologicals (Kolkata), Institute of Animal Health &Biologicals (Bangalore), Animal Health Institute (Jallandhar), and Institute of Veterinary Biologicals, Khanapara (Guwahati) are functioning as referral laboratories for Western, Eastern, Southern, Northern and North-Eastern regions, respectively. The laboratories at NRDDL (Jalandhar), SRDDL (Bangalore), ERDDL (Kolkata), and CDDL (Izatnagar) have been strengthened with pre-fabricated BSL-III laboratories while a mobile BSL-III laboratory has been provided to NERDDL, Guwahati. These RDDLs aids in the surveillance and diagnosis of various livestock and poultry diseases, including Avian Influenza and Bovine Spongiform Encephalopathy (BSE). under umbrella scheme "White Revolution - Rashtriya Pashudhan Vikas Yojana" as "Veterinary Services" and the funding pattern is 60:40 between Centre and State (90:10 for the North Eastern States and the Himalayan States and UT's 100%). The scheme comprises of the following subcomponents;

(a) Assistance to States for Control of Animal Diseases (ASCAD)

(b) Professional Efficiency Development (PED)

(c) National Project on Rinderpest Surveillance & Monitoring (NPRSM)

(d) Foot and Mouth Disease Control Programme (FMD-CP): *(This component is withdrawn from LH & DC from January 2020 as covered under* National Animal Disease Control Programme (NADCP))

(e) National Animal Disease Reporting System (NADRS)

(f) Peste des Petits Ruminants Control Programme (PPR-CP)

(g) Brucellosis Control Programme (Brucellosis - CP) (This component is withdrawn from LH & DC from January 2020 as covered under National Animal Disease Control Programme (NADCP))

(h) Establishment and Strengthening of existing Veterinary Hospitals and Dispensaries (ESVHD)

(i) Classical Swine Fever Control Programme (CSF-CP)

6.9. Dairy Schemes & Policies - Gujarat State

Gujarat State is a leading state for milk production in the country. "Amul" pattern is well known and accepted by all the states in our country and some of the other countries also. Milk procurement by co-operative movement is basic them in our state. At village level 19576 Milk Co-operative societies 102 chilling centres and 18 Dairy processing units at district level (Dairy) are in functioning in the state. State Government has given full support to the Dairy Development through Dairy co-operation movement. Ten District Co-operative Union has established Twelve Cattle Feed Factories to produce and supply cattle feed to their members at village level at no profit no loss basis. The total production of cattle feed is 2092683.50 M.T. by above nine factories. Twenty-Two Cooperative Dairy Unions have a total 250.4 Lakh Liter Per Day milk install capacity and they procured 237.03 LLPD milk. During the year 2018-19, these 22 dairy unions have 102 chilling centres also having a capacity of 72.42 LLPD of milk.

National Dairy Development Board Anand has made substantial investments in Junagadh Dairy, now they have starting milk-procurement activities in Junagadh, Porbandar, Amreli, Jamnagar, and Kutch Districts. Above all DCMPU collectively Installed capacity is 8.45 LLPD. All district of Kutch Saurashtra regions has started a co-operative dairy union. During the year 2018-2019 the Porbandar, Devbhumi Dwarka, Morbi, Botad, Junagadh, Jamnagar, Amreli, Bhavnagar, Rajkot, Surendranagar, and Kutch have collectively produced 27.75 LLPD.

Table: 6.9.1: Gujarat State Plan Dairy Development Scheme 2018 – 19

Sr.No.	Scheme	Physical Target for the Year 2018-19	Eligibility Criteria	Assistance to be given
1	Concentrate feeding for pregnant Animal (Cow + Buffalo) of Schedule tribe animal keeper	5000	Schedule tribe animal keeper	75% of purchase cost or maximum up to Rs.3000/-
2	Concentrate feeding for pregnant Animal (Cow+ Buffalo) of Schedule Cast animal keeper	2500	Schedule cast animal keeper	75% of purchase cost or maximum up to Rs.3000/-
3	Concentrate feeding for pregnant Animal (Cow + Buffalo) of General category animal keeper	10,000	All animal keepers	75% of purchase cost or maximum up to Rs.3000/-
4	Assistance for purchase of milking machine of schedule tribe animal keepers	150	Schedule tribe animal keeper who is keeping more than 05 milch animals	75% of purchase cost or maximum up to Rs.33,750/-

Sr.No.	Scheme	Physical Target for the Year 2018-19	Eligibility Criteria	Assistance to be given
5	Assistance for purchase of milking machine of schedule cast animal keepers	100	Schedule Cast animal keeper who is keeping more than 05 milch animals	75% of purchase cost or maximum up to Rs.33,750/-
6	Assistance for purchase of milking machine of General Category animal keepers	2000	All animal keeper who is keeping more than 05 milch animals	75% of purchase cost or maximum up to Rs.33,750/-
7	Assistance for Dudhghar Construction for the village Dairy co-operative societies of schedule tribe area of the state	12	Village Dairy cooperative Societies of tribe area of the state	50% of the Unit cost or maximum up to Rs.5,00,000/-
8	Assistance for Dudhghar Construction for the village Dairy co-operative societies of schedule cast area of the state	6	Village Dairy cooperative Societies of Schedule Cast of the state	50% of the Unit cost or maximum up to Rs.5,00,000/-
9	Assistance for Dudhghar Construction for the village Dairy co-operative societies of the state	100	All Village Dairy Co-operative Societies of the state	50% of the Unit cost or maximum up to Rs.5,00,000/-
10	Assistance for Godown Construction for the village Dairy cooperative societies of schedule tribe area of the state	8	Village Dairy cooperative Societies of tribe area of the state	50% of the Unit cost or maximum up to Rs.5,00,000/-
11	Assistance for Godown Construction for the village Dairy cooperative societies of schedule cast area of the state	4	Village Dairy cooperative Societies of Schedule Cast of the state	50% of the Unit cost or maximum up to Rs.5,00,000/-
12	Assistance for Godown Construction for the village Dairy co-operative societies of the state	100	All Village Dairy Co-operative Societies of the state	50% of the Unit cost or maximum up to Rs.5,00,000/-

Sr.No.	Scheme	Physical Target for the Year 2018-19	Eligibility Criteria	Assistance to be given
13	Assistance for the purchase of Automatic milk collection system (AMCS) for the village Dairy co-operative societies of tribe area of the state	50	Village Dairy cooperative Societies of tribe area of the state	80% of the Unit cost or maximum up to Rs.80,000/-
14	Assistance for the purchase of Automatic milk collection system (AMCS) for the village Dairy co-operative societies of schedule cast of the state	25	Village Dairy cooperative Societies of Schedule Cast of the state	80% of the Unit cost or maximum up to Rs.80,000/-
15	Assistance for the purchase of Automatic milk collection system (AMCS) for the village Dairy co-operative societies of the state	300	All Village Dairy Co-operative Societies of the state	80% of the Unit cost or maximum up to Rs.80,000/-
16	Assistance for the purchase of Bulk Milk Cooler (BMC) for the village Dairy cooperative societies of tribe area of the state	20	Village Dairy cooperative Societies of tribe area of the state	80% of the Unit cost (As per the capacity of the Bulk Milk Coolers)
17	Assistance for the purchase of Bulk Milk Cooler (BMC) for the village Dairy cooperative societies of schedule cast of the state	10	Village Dairy cooperative Societies of Schedule Cast of the state	80% of the Unit cost (As per the capacity of the Bulk Milk Coolers)
18	Assistance for the purchase of Bulk Milk Cooler (BMC) for the village Dairy co-operative societies of the state	75	All Village Dairy Co-operative Societies of the state	80% of the Unit cost (As per the capacity of the Bulk Milk Coolers)
19	Assistance for the purchase of Milk Adulteration Detection Machine (MADM) for the village Dairy cooperative societies of tribe area of the state	15	Village Dairy cooperative Societies of tribe area of the state	75% of the Unit cost for the Village Dairy co-operative societies of the Tribe Area

Sr.No.	Scheme	Physical Target for the Year 2018-19	Eligibility Criteria	Assistance to be given
20	Assistance for the purchase of Milk Adulteration Detection Machine (MADM)for the village Dairy cooperative societies of schedule cast of the state	10	Village Dairy cooperative Societies of Schedule Cast of the state	75% of the Unit cost for the Village Dairy co-operative societies of the Tribe Area
21	Assistance for the purchase of Milk Adulteration Detection Machine (MADM) for the village Dairy co-operative societies of the state	50	All Village Dairy Co-operative Societies of the state	50% of the Unit cost for Village Dairy co-operative societies of the state

Source: DOAH, 2020

7

Prospects and Challenges

7.1. Challenges in Milk Quality

Demand for milk and milk products in India is growing very rapidly, owing to urbanization, customer convenience, and the change from loose to packaged milk products. At present, the dairy sector is rising at about 10-12 per cent annually. Based on population growth projections and an increase in urbanization over the next four decades, India is projected to need approximately 600 million tonnes of milk per year to meet the demand for milk and milk products. This means that India's milk production needs to grow at around 3.2 per cent CAGR for the next 40 years (FICCI, 2020).

Main reason of poor quantity of milk is due to unhygienic and poor condition of animal farms and dairies. Even due to high demand in foreign country we cannot export such poor quality of milk product. Many times, Indian milk products get rejected in foreign nations due to unhygienic milking milk contains high number of microbial counts. Reasons behind the quality affected in India are due to poor health of animal, polluted water and food and Unclean surrounding in farm. So, it is necessary to consider all this challenges in improving milk production. Currently most of the milking animals are not screened on regular interval of time. So, there is lots of chances that some diseases can transmitted to human beings. Farmers use antibiotics and chemical drugs to increase the production of milk this will also affect the quality of milk. These issues can be avoided by regular screening of animals in animal husbandry and also maintaining history records of health, feed etc. Quality of feed should be checked regularly. There must be pollution control in the feed and water used in cattle farms (Bhagyalaxmi, 2020).

While the Food Safety and Standards Authority of India (FSSAI) has laid down standards of quality for milk and milk products, there are still concerns on the available products and their acceptability towards adhering to standards. As per the latest survey by FSSAI, around 90 per cent of products are found safe. Nevertheless, there remain doubts in the minds of consumers on the quality aspect (FICCI, 2020). A major finding in the survey was the presence of Aflatoxin M1 residues beyond permissible limits in 368 (out of 6,432) samples, that is 5.7 per cent of the samples. The survey further showed that 77 out of 6,432 samples, that is 1.2 per cent of the samples, had residues of antibiotics above the permissible limits. Due to Underdeveloped Facilities for Quality and Hygienic Livestock Management, despite being home to the largest livestock population in the world, India's livestock management is below par with other countries. There is a shortage of organized space and shelter for horde milk-producing animals. The veterinary infrastructure is also poorly developed.

7.2. Challenges in Milk Productivity

According to NITI Aayog's working paper on demand and supply projections towards 2033, the positive growth in the bovine population has contributed to the significant increase in milk production in the country. A variety of constraints are exposed to the livestock industry. The pre-dominant is low productivity, chronic shortages of feed & fodder, the large population of unproductive cattle, absence of effective extension system, low health care, immunization, and hygienic program, lack of cold chain logistics, unorganized marketing, etc.

Another challenging area for dairy development has been that while the availability of fodder (dry) and feed has increased significantly in recent years, the increase in green forages has been negligible. The deficit of dry fodder concentrates and green fodder is estimated at 10 per cent, 33 per cent, and 35 per cent, respectively. This deficit is likely to widen to 11 per cent, 35 per cent, and 45 per cent, respectively by 2020-21. Another critical issue faced is the lack of data on the availability and production of various types of green fodder. The Ministry of Agriculture does not collect data on fodder crops relative to food crops, the availability of which poses a major challenge to increase productivity.

Management of the major feeding sources for livestock is the main area of concern. They include pasture lands, crop residues, edible weeds, grasses, cultivated fodder, tree leaves, and agro-industrial by-products. Fodder crops are cultivated only on about 4.9 per cent of the gross cropped area of the country and this area has been static for the last 25 years. Thus, the government has to formulate a contingency plan to boost fodder production so that the growth in milk production is sustained in the next few decades as well.

7.3. Challenges in Milk Demand, Supply, and Dairy Cold Chain

Although FSSAI has registration and licensing data on the ability of dairy plants in the dairy sector, it is not subdivided into private and cooperative sectors. In the absence of data on the processing capacity of the private sector, it is challenging to formulate policy supporting private sector participation in the growth of the dairy sector. While the data about the milk processing capacity of the dairy cooperatives is regularly updated by NDDB, the processing capacity added by the private sector has witnessed significant growth and expansion in the last two decades. But it has yet to be captured via information. In collaboration with FSSAI, the DAHD can employ a private entity to compile data on the milk processing ability of the private sector. As NDDB does not keep data on private sector dairy players, suitable policy changes cannot be brought about in specific states where cooperatives are weak. The task of collection and compilation of data about the private sector processing capacity is therefore very important.

Fluctuations in demand and milk prices should be addressed on an urgent basis as it impacts the farmers' income. It also discourages farmers from growing their herd size, as incomes are volatile and risky. There should be a capacity enhancement for processing and storing excess milk as Skimmer Milk Powder (SMP). India can explore the possibility of exports of SMP especially in the vicinity, i.e., South Asian countries.

7.4. Constraints in Milk Marketing

The dairy sector is characterized by small, dispersed, and unorganized holders of milk-animals.; Low productivity; poor and inadequate feeding and health care for animals; lack of a guaranteed year-round remunerative producer price for milk; Insufficient basic infrastructure for the procurement of manufacturing materials and services; Inadequate basic milk procurement, distribution, manufacturing, and marketing facilities; and lack of professional management. The prevalence of mixed crop-livestock farms and the fact that most of the milk animals are fed on crop by-products and residues are other significant features of the dairy sector., Which have very low costs of opportunity. Furthermore, the policies and programs implemented for milk production, including those relating to foreign trade, are not conducive to promoting sustainable and equitable dairy development.

A significant restriction to milk production is the low productivity of milk animals. By crossbreeding low-yielding nondescript cows with high-yielding selected indigenous purebreds or appropriate exotic breeds in a phased manner, the productivity of dairy animals could be increased. In addition to emphasizing milk yield, the cattle-breeding strategy should also provide for the production of good-quality bullocks to meet the draft power requirements for agriculture. In all areas where buffalo are well-adapted to agro-climatic conditions, high priority should be given to upgrading nondescript buffalo through selective breeding with high-yielding purebreds such as Murrah, Mehsani, or Nili Ravi.

Producers' interests should receive the utmost consideration when setting procurement rates. At least the long-term average cost of milk production should cover the producer price and provide a decent mark-up. The Departments of Animal Husbandry or the Dairy Development Boards/Corporations should undertake studies of milk production costs and their financial viability. Such studies should be carried out in all major agroclimatic zones and should be replicated at regular intervals of approximately three years to assess whether milk production is profitable and provide an objective basis for evaluating the milk price of the producer.

The dairy sector holds high promise as a secure source of livelihood for the vast majority of the rural poor, considering all the problems it faces. Under the current WTO trade regime, the liberalization of world trade in milk products presents new challenges and has opened up the new export potential for the dairy industry. India's dairy sector needs to improve its competitive economic advantage in milk products, both in terms of quality and cost, and in terms of its reputation in international markets. The government's role should be to direct, organize and regulate the activities of various dairy development organizations; to establish and maintain a level playing field for all stakeholders; and to create and maintain a healthy socio-economic, institutional and political ecosystem for the development of small dairy farmers.

It is important to formulate a detailed milk production strategy. Such a strategy should be an integral part of the national development policy and its direct and indirect effects on other sub-sectors of the economy should be properly taken into account and vice versa. Continued adaptation of management strategies to suit markets, ecosystems, and socio-economic conditions would also depend on the future of dairy production.

7.5. Challenges in Exports

The majority of the large dairy importing nations across the world are not permitting the import of dairy from India by applying various tariff and non-tariff measures to protect imports. A broad list of measures imposed by various countries on dairy imports is as below:

Table 7.1: List of Trade barriers imposed by countries on India

Country	**Challenges Faced by Indian Dairy Exporters due to Restrictions**
China	The world's largest importer of dairy products does not permit import from India due to a reciprocal ban.
Russia	Requires their veterinary control team to approve India's plants. Since the last 4 years no approval for the Indian dairy industry

European Union	Not approved any Indian dairy plant (for exports) under the pretext of veterinary control, Antibiotic and Pesticide residue, etc.
Canada	Canadian Food Inspection Agency does not permit the import of dairy products from India. Impose Import duty of around 250%.
Indonesia	Requires MUI (Halal) as well as Agricultural department approval. MUI team visits for plant inspection.
Malaysia	Requires JAKIM (Halal) approval. Visits for plant inspection once in 2-3 years and inspects very limited plants.
USA	The import duty of 40 to 60% on dairy products
South Africa, Mexico, Venezuela	Do not allow the import of dairy products from India.
Australia	Do not permit the import of dairy products from India due to Foot and Mouth Disease (FMD). Only retorted products are permitted
Egypt / Saudi Arabia	Follows EU standard for Aflatoxin and stopped Indian milk powder.
Pakistan	45 % import duty on SMP and does not permit the import of Other dairy products
Sri Lanka and Bangladesh	30% import duty on dairy products
Thailand	More than 40% import duty
Myanmar and the Philippines	The long registration process for dairy imports
Afghanistan	Though large importer, the goods have to be routed via Karachi (because banks not permitting Iran route)

Source: FICCI, 2020

Developed countries such as the EU and the USA have provided their dairy farmers with tremendous subsidies for exports. According to the latest figures, the US is providing subsidies of over USD 12 billion and Europe is providing support from time to time to its dairy farmers of about USD 15 billion under export subsidies and assistance programs. Besides, developing countries provide other supports by providing buffer stocking storage support, or additional market. In India, there is no such support provided to dairy farmers thus making them uncompetitive in global trade.

Inclusion of Dairy Products under New Freight Support Scheme

As part of the 2018 Agriculture Export Policy Implementation Action Plan, the Government announced the scheme for providing financial assistance for the transport and marketing of agricultural products to improve exports of such commodities to some European and North American countries. The current policy for providing the Transport and Marketing Assistance Scheme (TMA) as per Public Notice No 82/2015.2020 dated 29/03/2019 provides a boost for the export of specified agricultural products with the dairy sector is excluded.

Because of the enormous potential of the dairy industry in terms of growing the export basket of India, the government shall look forward to expanding this scheme to dairy products as well.

7.6. Opportunities in Dairy Sector

With the series of measures being taken by the government as well as the growing role of the private sector in dairy development, India is expected to sustain its growth in milk production and milk processing in the coming decades. However, the critical issue facing the dairy sector is how to bring down the cost of milk production through investment to boost productivity and also explore possibilities of increasing dairy products exports. In addition to increased processing capacities, the future of the Indian dairy industry lies in making its raw material supplier, the farmer, a cost-effective and globally competitive producer. The role of the private sector in developing an efficient dairy value would be critical, as India aims at doubling farmers' income in the next few years.

- Soft loans, the moderate interest rate on CAPEX, and working capital support
- Subsidies, incentive, and other financial support at par with cooperatives
- NDDB's operations should not be restricted to the private sector alone. It should work for the development of the entire dairy sector, rather than only the cooperative sector.
- National Cooperative Dairy Federation of India (NCDFI) manages an e-marketing portal (www.ncdfiemarket.com) and allows private dairies to buy milk and products but prohibits them from selling their dairy produce on their electronic platform
- Setting up an Institutional structure and corpus fund to purchase milk powders (SMP) and butter when excesses are stocked
- Incentivize exports – International prices of dairy commodities are often lower than the production cost in India.

Source: FICCI, 2020

7.7. Opportunities for Dairy Exports

As per the USDA estimates, out of India's total production of 190 MMT, nearly 55 per cent milk (105 MMT) originates from buffalos which have an average milk yield of 6.2 Kg per day (as per the latest DAHD's report). Water buffalo milk has a higher fat content, which contributes to higher consumer prices, as the price of milk is measured by the volume, fat, and solid non-fat (SNF) content. Certain products like Buffalo Mozzarella Cheese, Buffalo Ghee, etc. can command a huge premium in markets like the United States and Europe and are also preferred by Dairy farmers due to high returns.

India could investigate the possibility of indigenous Indian dairy-based sweets like Gulabjamun, Rosgulla, Penda, etc. The Government could support exports through the provision of technical and financial support for the manufacture of such goods since they are mainly handmade. The government can allocate separate funds for dedicated marketing of value-added, Geographical Indication Registered, Region-specific produces. Both digital and traditional media platforms can be utilized for the sustained communication campaign.

Merchandise Exports from India Scheme (MEIS) support is currently provided at 3 per cent to Ice-cream and dairy-based ethnic sweet (Gulab jamun, Rasmalai, Rasgulla, Milk cake, etc.) and 5 per cent to products like Ghee, Butter, Cheese. MEIS support of 20 per cent on all dairy products including Milk powder, Tetra Pak Milk, Fermented Milk, and milk fat was given for a limited period from July 2018 to Jan 2019. Subsequently, India witnessed a substantial increase in the export of dairy products from the country and become competitive in the international market with this support. To boost the exports of dairy products, the Government should provide support in form of either MEIS or Rebate of State and Central Levies and Taxes (RoSCTL) to all dairy products.

As India had taken a firm stand in RCEP (Regional Comprehensive Economic Partnership) considering the impact on dairy farmers due to fear of import of dairy products from Australia and New Zealand, the government must continue the same policy. The government must ensure that none of them (Australia & New Zealand) are granted access in the coming Free Trade Agreements (FTA) or Comprehensive Economic Partnership Agreement (CEPA) with countries such as EU & USA etc as it would have a very bad effect on the livelihoods of millions of dairy farmers and the 'Atma-Nirbhar Bharat' status of India for milk products.

There is a huge potential for increasing Dairy Exports from India which can be explored by providing proper financial incentives, lower cost of doing business, and market support to boost exports. Further, a Stable policy framework from the Government with due impetus on Dairy Sector will allow all stakeholders to build business models with medium- to long-term plans, creating a strong Unique Selling Proposition (USP) for an Indian dairy export product.

8

Conclusion

a) **A Key GDP Contributor:** Dairy and Livestock sector form the backbone of agriculture with 28 per cent contribution in GDP and the progress in this sector is dependent on innovative measures to increase the value for farmers as well as customers.

b) **Doubling Farmer's Income :** Doubling Farmer's Income by 2022 can be achieved by improving crop and livestock productivity, efficiency to utilize resources, increased intensity of cultivation and diversification in high value cash crops, processing and value addition of food and animal products. This price realization can be assured through market reforms and efficient market linkages.

c) **Unorganised sector – a challenge :** The Indian dairy sector is still characterized by strong presence of unorganized sector. Apart from the Cooperative dairy sector, in the last few years there has been a considerable growth in Private Dairy sector and Milk Producers Organization like Maahi in Gujarat. If small dairy farmers are provided with sufficient 'value addition' training and support, they can increase their income considerably.

d) **Value addition :** Further, the milk processing plants can also go for value addition and innovation and in turn pay remunerative prices to milk producers.

e) **Infrastructural facility:** The infrastructural facility for strengthening milk handling capacity of the organized sector can be increased from 30 per cent to 50 per cent of the total milk production (Bhandari, 2020). This will help in producing quality milk products and even export these products.

f) **FPOs – a key player :** Marketing of the surplus milk and value-added products can be effectively done with the help of Farmer Producer Organization especially for small and marginal dairy farmers who keep only few animals and lack the bargaining power. Collective pooling of raw material along with shared resources can help individual farmers who are not associated with co-operatives in achieving economies of scale and adopting standardized practices.

g) **Risk Management :** During the difficult times like lockdown days due to COVID-19, Indian dairy Industry suffered 25 to 30 per cent dip in the demand as out of home consumption like restaurants, eatery, tea stall, etc were completely shut down. However, even in this situation milk procurement remained constant due to Government directives of preserving farmers' livelihood. This resulted into huge surplus of milk in most states.

h) **Storage Facility – a key factor** : Co-operatives who were not well equipped to store and process the surplus like in the states of West Bengal and Jharkhand were forced to throw of the milk. The other Co-operatives were making white butter and SMP from this surplus. The situation so changed that the price of SMP which was Rs 340 per kg prior to COVID-19 dipped to Rs 230 per kg as the global demand dipped.

i) **Innovation and role Model :** AMUL is a live example of how efficient management and timely innovation can increase the sales in these difficult times where private dairies were struggling to survive.

References

APEDA (2020). India Export Statistics. Retrieved from http://agriexchange.apeda.gov.in/indexp/genReport_combined.aspx

APEDA. (2020). India production of Milk. Retrieved from http://agriexchange.apeda. gov.in/India%20Production/AgriIndia_Productions.aspx?productcode=1023

Bhagyalakshmi, M (2020). A Study on Major Issues and Challenges of Dairy Farmers in India; Science Technology and Development Journal, 9(4), (pp. 166 – 172).

Bhandari, G. (2020). Implications of COVID-19 for Indian Dairy Industry. Food and Scientific Reports. Retrieved from https://www.researchgate.net/ publication/343253865_Implications_of_COVID-19_for_Indian_Dairy_Sector

Brindha, N., Chitra, P., Janarthanan, R., & Murali, A. (2017). A Study on Detection of Adulteration in Milk Samples from Different Regions of Thuraiyur District in Tamil Nadu, India. Int. J. Curr. Microbiol. App. Sci, 6(12), (pp. 3303-3310).

Compassion in World Farming (CIWF, 2012). Statistics: Dairy Cows. (p.2).

Confederation of Indian Industries (CII). (2020). Impact of Covid-19 on the Indian dairy segment [Press release]. Retrieved from http://face-cii.in/sites/default/files/dairy_press_release_final.pdf

Dairy - India. (2018). State Wise List of Dairy Plants. Retrieved December 15, 2020, from https://www.ruralrelations.com/wp-content/uploads/2018/09/State-Wise-Listof-Dairy-Plants.pdf

Department of Animal Husbandry & Dairying (DAHD). (2017). Basic Animal Husbandry & Fisheries Statistics 2017 (pp. 20 - 26).

Department of Animal Husbandry & Dairying (DAHD). (2019). Retrieved 23 November 2020, from https://dahd.nic.in/about-us/divisions/cattle-and-dairy-development

Department of Animal Husbandry and Dairying. (2019). Annual Report 2019-20 (pp. 10, 12 - 101).

DGCIS Principal commodities data April - March (2019-20). Comparative statement for export of agri and processed food products April - March (2019-20) [Dataset].

Retrieved from https://apeda.gov.in/apedawebsite/Comparative_Statement Mar_2019- 20.pdf

Directorate of animal husbandry (DOAH). (2019). 36th Survey Report on Estimates of Major Livestock Products for The Year 2018-2019 Gujarat state (pp. 13-72).

Directorate of animal husbandry (DOAH). (2020). Retrieved 3 December 2020, from https://doah.gujarat.gov.in/dairy-development.htm

FAO (2018). Gateway to dairy production and products. Food and Agriculture Organization of the United Nations, Rome.

Federation of Indian Chambers of Commerce & Industry (FICCI). (2020). Development of Dairy Sector in India (pp. 34 - 38).

Fernandes, N. (2020). Dairy industry witnesses an innovation galore during Covid times. Dairy Times, 4(5). Retrieved from https://agronfoodprocessing.com

Food Safety and Standards Authority of India (FSSAI). (2018). National Milk Safety & Quality Survey (pp. 10 - 14).

Hari, H., Shree, N., & Basak, D. (2020). Impact of Covid-19 on Agriculture in India. Journal of Economics & Policy Analysis, 1(1), 82-94.

Indiaagronet. (2012). Summary of utilization of milk in India. Retrieved December. Retrieved from http://www.indiaagronet.com/indiaagronet/Dairy/totalmilkproduction.htm

International Dairy Federation (IDF). (2020). Annual Report 2019 - 20. (p.5 – 12).

Joshi, V. (2012). Econometric Analysis of the India–Sri Lanka Free Trade Agreement. Asian Economic Journal, 26(2), 159-180.

Kumar, A., Mishra, A. K., Parappurathu, S., & Jha, G. K. (2018). Farmers' choice of milkmarketing channels in India. Economic & Political Weekly, 53(51), (pp. 58-67).

Ministry of Statistics & Programme Implementation (MOSPI). (2012). India - Household Consumer Expenditure, NSS 68th Round, (pp. 24-25)

Ministry of Statistics and Programme Implementation (MOSPI). (2017). Key Indicators of Household Social Consumption (pp. 30-32).

Mondal, P. (2012). Comparison of Dairy Farming in India with Advanced Centres. Retrieved December 19, 2020, from https://www.yourarticlelibrary.com/dairy-farmmanagement/comparison-of-dairy-farming-in-india-with-advanced-centres/35767 National Dairy Development Board (NDDB). (2020). Annual Report 2018 - 2019 (p. 29).

National Dairy Development Board (NDDB). (2020). National Dairy Plan - I Brief. Retrieved from https://www.nddb.coop/ndpi/about/brief

NITI Ayog. (2020). Demand & Supply Projections towards 2033 (pp. 124 - 125).

Office of Economic Adviser. (2020). Retrieved 20 December 2020, from https://eaindustry. nic. in/display_data_201112.asp

Office of Economic Adviser. (2020). Retrieved 23 November 2020, from https://eaindustry. nic. in/display_data_201112.asp

Ohlan, R. (2014). Globalization and the dairy industry. Stadium Press (India) Pvt. Ltd, New Delhi.

Palanisingh, V., Salahudeen, V. A., & Gurumoorthy, T. R. (2017). Dairy exports in India: Production and consumptions. International Journal of Advanced Research in Management and Social Sciences, 6(5), (pp. 70-75).

Parida, Y., Ghule, A. K., & Dudharejiya, P. (2019). Trade competitiveness of Indian dairy industry: an empirical analysis. (pp. 4 – 10).

Rajendran, K., & Mohanty, S. (2004). Dairy co-operatives and milk marketing in India: Constraints and opportunities. Journal of Food Distribution Research, 35(8562016-56967), (pp. 34-41).

Rao, K. H., Raju, P. N., Reddy, G. P., & Hussain, S. A. (2013). Public-private partnership and value addition: A two-pronged approach for sustainable dairy supply chain management. IUP Journal of Supply Chain Management, 10(1), (pp. 15).

Sabnavis, M., Kansara, D., & Bagariya, R. (2020). Indian Dairy & Dairy Products Industry Update [Ebook] (pp. 4 - 10). CARE Ratings. Retrieved from https:// www.careratings. com/upload/NewsFiles/SplAnalysis/Indian%20Dairy%20%20 Dairy%20Products%20 Industry%20-%20June%202020.pdf

Salois, M. (2016). Global dairy trade situation and outlook. International Food and Agribusiness Management Review, 19(1030-2016-83105), (pp. 11-26).

Singh, S. (2015). Promotion of Milk Producer Companies: Experience of NDS. In Presentation in Workshop on Producer Companies organized by National Dairy Development Board, Anand, Gujarat (Vol. 8) (pp. 12).

Singh, S. R., & Datta, K. K. (2010). Understanding value addition in Indian dairy sector: some perspectives. Agricultural Economics Research Review, 23(conf), (pp. 487493.).

United States Department of Agriculture & Global agricultural Information Network (USDA GAIN, 2020). Dairy and Products Annual (p. 3, 7-10).

United States Department of Agriculture & Global agricultural Information Network (USDA GAIN, 2020). Budget Raises Tariffs on Several Agricultural Products (pp. 3-12).

V. K. Shilby & H. N. Mishra (2013). Fermented Milks and Milk Products as Functional Foods—A Review, Critical Reviews in Food Science and Nutrition, 53:5, (pp. 482496), DOI: 10.1080/10408398.2010.547398.

Websites Used:

https://www.amul.com/ https://aavinmilk.com/ https://www.steragro.com/ https://www.verka.coop/ https://www.creamlinedairy.com/ http://britannia.co.in/products/dairy https://www.kwality.com/ https://vadilalgroup.com/ https://www.nestle.com/brands/dairy https://www.motherdairy.com/ http://www.schreiberdynamixdairy.com/ https://haryanamilk.com/

Index